IMMUNITY

HOW ELIE METCHNIKOFF CHANGED THE COURSE OF MODERN MEDICINE

LUBA VIKHANSKI

First edition
Published by Chicago Review Press Incorporated
814 North Franklin Street
Chicago, Illinois 60610
ISBN 978-1-61373-110-9

Library of Congress Cataloging-in-Publication Data
Names: Vikhanski, Luba, author.
Title: Immunity : how Elie Metchnikoff changed the course of modern medicine
/ Luba Vikhanski.
Other titles: How Elie Metchnikoff changed the course of modern medicine
Description: First edition. | Chicago, Illinois : Chicago Review Press Incorporated, [2016] | Includes bibliographical references and index.
Identifiers: LCCN 2015050225 | ISBN 9781613731109 (alk. paper)
Subjects: | MESH: Metchnikoff, Elie, 1845-1916. | Allergy and Immunology | Microbiology | Immune System Phenomena | Ukraine | France | Biography
Classification: LCC RC584 | NLM WZ 100 | DDC 616.97–dc23 LC record available at http://lccn.loc.gov/2015050225

Typesetting: Nord Compo

Printed in the United States of America
5 4 3 2 1

To my mother and the memory of my father

The scientist must concern himself with what one will say about him in a century, rather than with current insults or compliments.

—Louis Pasteur, in a letter to Adolphe Guéroult, editor in chief of *L'Opinion Nationale*, July 19, 1864

CONTENTS

IV Not by Yogurt Alone

V Legacy

A NOTE ABOUT NAMES

THE VARIOUS NAMES ILYA ILYICH METCHNIKOFF used outside his native Russia reflect the adjustments people inevitably make when reaching out to foreign audiences or adapting themselves to new countries.

When he published his papers abroad, and later when he left Russia, he changed his first name, Ilya, to that by which his namesake prophet—Elijah in the English-speaking world—went in each particular country: Elias in Germany and Elie in France. He wrote his last name so that it would be pronounced as close as possible to the original Russian: Mecznikow, Metschnikoff, and finally Metchnikoff, as spelled in France, where he ultimately settled. Mechnikov, the standard transliteration of his last name, reflects its Russian spelling but not the pronunciation; the *-ov* endings of Russian names sound more like *-off.*

The numerous variations have occasionally resulted in a lack of consistency even within a single organization. His Nobel Prize diploma, for example, states his name as Elie Metchnikoff, but the official Nobel Prize website lists him as Ilya Mechnikov.

I refer to him as Ilya during his earlier years, but when writing about his later life, I've chosen to use the first and last names he ultimately adopted outside Russia: Elie Metchnikoff. In France, his first name is spelled Élie, but in English-language publications, it has always been spelled without the acute accent, as it is in this book.

I

MY METCHNIKOFF

1

REVERSAL OF FORTUNE

ON JULY 15, 1916, THE WEATHER in Paris was overcast and oppressively humid. Bleak light poured through the metal-framed windows of Louis Pasteur's former apartment upon gold-patterned wallpaper, oriental rugs, and carved antique furniture. The museum-like residence at the Pasteur Institute was filled with oil paintings, vases, statuettes, and other works of art Pasteur had received as gifts from grateful admirers. As Elie Metchnikoff lay in this shrine of science, pillows propping his large head with its mane of gray hair and beard, he held the hand of his wife Olga, fifty-seven, a slim, oval-faced blonde sitting at his bedside. He had devoted his entire life to science. Now science was letting him down.

Metchnikoff knew he was dying, but his worst fear was not death itself. What he dreaded most was that his passing away at seventy-one, decades too early by his own standards, would discredit his theories about life, health, and longevity.

He had been much better at creating new areas of research than at fitting into existing ones. The 1908 Nobel Prize in Physiology or Medicine had been his reward for helping found the modern science of immunity. He had then launched the first systematic study of aging, coining the term *gerontology*. In future generations, he argued, people could live to 150. To stay healthy, he believed they had to repopulate their intestines with beneficial microbes to replace harmful ones—for instance, by eating yogurt or other forms of sour milk.

Coming from such an acclaimed scientist, these ideas had created a sensation, making him an international celebrity and turning sour milk into a global mania. In a 1911 poll by a British magazine, he had been voted one of the ten greatest men in the world. But now that his heart was failing him less than halfway to his own target, his teachings threatened to die with him.

"Remember your promise—you'll perform my autopsy," he told an Italian physician who entered the room, one of his numerous trainees at the Pasteur Institute, where Metchnikoff had worked for nearly three decades. "And pay attention to my intestines, I think there's something there." Even after his death, he hoped to be doing what he had done all his life: serving as his own subject for research. When he made an abrupt movement, Olga pleaded with him to lie still. He didn't answer; his head had fallen back on the pillow.

The tricolor national flag on the Pasteur Institute's facade was lowered to half-mast, draped in black.

Millions of people were dying in a world war that had been raging for nearly two years, but this particular death was major news around the globe. The international press was unanimous in praising Metchnikoff, placing him beside Pasteur, Lord Lister, and Robert Koch among "the immortals in the lifesaving science of bacteriology." But just as he had feared, his death delivered a fatal blow to his theories of longevity.

Soon afterward, his name sank into oblivion. Hardly anyone followed up on his research, neither in aging nor in immunity. Only the yogurt craze he had launched on both sides of the Atlantic proved immortal. Indeed, outside the Pasteur Institute, to the extent that he was remembered at all after his death—except by historically minded immunologists and a handful of life-extension enthusiasts—it was usually in connection with yogurt.

There was one exception. In his homeland, Ilya Ilyich Metchnikoff was not just remembered but revered, and not at all because of yogurt.

When I was a girl growing up in Moscow in the early 1970s, Metchnikoff was known as the Russian Pasteur and upheld as a shining example of national talent. In company with Russia's other historic heroes, he was a cult figure, glorified by the Soviet regime as a way of instilling in the population a sense of belonging to a great nation. In my ninth-grade textbook, the ideologically weighted *History of the USSR*, he was canonized as "a model of selfless service to the Fatherland and to science." Like all the children in Moscow's Secondary School No. 732, to say nothing of the rest of the fifty million or so schoolchildren in the Soviet Union, I learned that Metchnikoff's valiant struggles against "bourgeois ethos" had helped pave the way for "Leninism, the highest achievement of Russian and world culture in the imperialist era."

But like many children of dissidents (my family had long-standing scores to settle with the repressive Soviet regime), I loathed this hero worship. In fact, I secretly suspected that Metchnikoff was a fake, together with most other Russian greats in my textbook, their accomplishments primarily products of communist propaganda. When I left Russia for good at seventeen, I would have been quite happy never to hear about any of them again.

Thirty years passed. While working as a science writer in Israel in the mid-2000s, I received an e-mail from Leslie Brent, a distinguished professor of immunology in the United Kingdom. In a search I had undertaken for little-known but key episodes in the history of science, I had asked him to list great immunologists whose life stories, in his opinion, had yet to be properly told. Metchnikoff's name, high on Professor Brent's list, jumped out at me. It brought back memories of my teenage self with my braids, fountain pens, and brown wool high school uniform. I was shocked by my own myopia. Had I wrongly dismissed a genius in my overall rejection of the party line?

To clear up this possible error of judgment that had been trailing me since childhood, I decided to investigate. Who was this man? Was he really as great as my textbooks had proclaimed him to be?

Delving into accounts of Metchnikoff's life, I discovered a tale worthy of becoming a legend. It was the story of a boy who grew up in an obscure village, dreamed about creating a theory that would revolutionize medicine—and went on to author the modern concept of immunity as an inner curative power. Today we take it for granted that our immune system protects us from within; it is hard to imagine that just over a century ago, mainstream medicine had no notion of the body's inner defenses. Then one day in the early 1880s, Metchnikoff straightened his spectacles, peered into his microscope, and declared that he was watching a curative force in action.

He was an outsider—a zoologist, not a physician. His theory of immunity came under attack in Germany; a prominent French scientist dubbed it "an oriental fairy tale." How did Metchnikoff become one of the founding fathers of immunology? And how did it happen that a scientist who launched such a radical shift in human consciousness was so thoroughly forgotten outside Russia?

Searching for answers, I started out with the adoring biography by his wife Olga, *Life of Elie Metchnikoff,* as well as books and articles written about him in Russia. But I wanted to form my own opinion. I sought out Metchnikoff's memoirs and letters and those by his friends and foes. I followed his trail to Ukraine, riding for hours on a vintage bus to Mechnikovo, the pastoral village where he had grown up, which was renamed in his memory. At a library during a trip to Moscow, I scrolled through the May 1945 microfilm of *Pravda,* which reported on the celebrations of the centenary of his birth. From the State Archive of the Russian Federation, I obtained reports filed on him by the tsar's secret police.

My initial intent was to write a book about an unjustly forgotten scientist. Then something unexpected happened: Metchnikoff's luck, I realized, had suddenly changed. In the 2000s, during the years in which I was rediscovering him, world science was rediscovering him as well. His ideas are now making a surprising comeback. He fought to leave his mark on the science of the twentieth century; instead, he advanced the science of the twenty-first.

Shifting my quest from the past to the present, I sought to understand this unusual reversal of fortune. What is Metchnikoff's true legacy? Is his research helping to rid humanity of disease, as he had hoped? And what does modern science say about gut microbes and longevity?

Throughout my searches, I wondered whether Metchnikoff himself would have felt vindicated by his own remarkable revival. Would he conclude that the fears he had on his deathbed were unfounded? And would he see himself as a winner in his major quests? If so, what a sweet victory it would be, with a tinge of vinegar for having been so long in coming.

My motivation to write a book about Metchnikoff only grew in the wake of his renewed relevance, but I was missing a crucial source of information.

2

THE PARIS OBSESSION

When passing through Paris in the fall of 2007, I looked for people who, in the distant past, might have heard firsthand stories about Metchnikoff. That was how I met Professor Elie Wollman, who had actually known Olga Metchnikoff. This encounter was to launch me on a Holmesian investigation that took an unexpected direction.

Wollman's father, Eugène, had been one of the last of Metchnikoff's students at the Pasteur Institute. Eugène gave his son, born after Metchnikoff's death, Metchnikoff's first name. The younger Wollman subsequently spent his entire career at the Pasteur Institute, where he made groundbreaking discoveries in microbial genetics and, like Metchnikoff, served as the institute's deputy director. Witty, vigorous, and sharp, Wollman (ninety years old at the time of our meeting) affectionately shared his memories of the widowed Olga, a family friend, who lived till 1944. I concentrated on every word. This was the closest I was going to get to someone who knew Metchnikoff.

"Olga was very beautiful, very fine. She possessed a great innocence," recalled Wollman—speaking French, he used the word *candeur*, which could also mean "ingenuousness" or "purity." "She kept the spirit of a young girl all her life; even her voice was girlish. I never heard her say anything unpleasant or nasty about anyone."

Olga's *candeur* evidently included relentless naïveté, a trait she shared with her husband. During World War II, when the Nazis were arresting Jews throughout occupied Paris, Wollman's parents went into hiding at the Pasteur Institute hospital, disguised as patients. In 1943, however, they were denounced and sent to a concentration camp.

"You know, in a way, I'm relieved. Your mother has been so tired lately, at least in the camp she'll be able to rest," Olga consoled the young Wollman with shocking blindness.

Both Eugène and Elisabeth Wollman died in Auschwitz.

As I was about to leave Wollman's elegant apartment, pausing to admire a pensive landscape painting by Olga in the living room, he dropped a clue: "Find out more about Lili Rémy." The name was familiar. Metchnikoff had served as Lili's godfather, as he had for the children of numerous other friends and colleagues. Lili was the daughter of Émile Rémy, the Pasteur Institute's scientific illustrator.

Something about the way Wollman mentioned Lili made me heed his advice. From other people versed in the institute's unofficial history, I learned that Lili had been "widely regarded" as Metchnikoff's out-of-wedlock daughter, or what the French call *un enfant naturel*, an expression hinting at tacit social acceptance.

I also learned that Lili had already died, but that her only son, Jacques Saada, had stayed in touch with the Pasteur Institute for a while, attending ceremonies and memorial events. Apparently unaware that his mother's alleged lineage was an open secret among Pasteur old-timers, he shrouded it in self-indulgent mystery. "He told it to you like a secret he hoped would not be kept," recalled a one-time staff member.

When I returned to Israel from Paris, finding Metchnikoff's only potential descendant—the son of his supposed love child—became imperative. Since Jacques Saada no longer lived at the address in the suburban town of Ville d'Avray that I had been given, I embarked on an inquiry. I called his former neighbors, officials in the Ville d'Avray municipality, and local real estate agents who might have sold his apartment. I even contacted lawyers in the neighborhood on the odd chance they knew him, since on the Pasteur Institute mailing list, Saada appeared as *Maître*, a French title for people in the legal profession.

Alas, I learned from Saada's former concierge that he had died several years earlier, but I kept searching. Perhaps I could contact his children? The prospect of finding Metchnikoff's direct descendants fired my imagination.

Finally, after months of long-distance phone calls and fruitless web surfing, I had a hit: I found Saada's name among the condolences in a 2003 issue of the online newsletter *Le Sévrien.* I obtained Saada's death certificate, which gave me the name of the person who reported his death: Dr. Patrice Rambert. Before I could finish introducing myself on the phone, Rambert exclaimed, "You must be calling about the Metchnikoff collection!"

I was dumbfounded: what collection? It sounded like a mystery plot, yet it was real. Metchnikoff's numerous objects and letters, Rambert told me, were locked up in a bank, on the Champs-Elysées of all places, and no one could take them out.

It turned out that before Saada died in 2003, he had placed his Metchnikoff collection in four safe deposit boxes at the Crédit Lyonnais bank on the Champs-Elysées. Under French law, anyone claiming a right to Saada's belongings would automatically inherit not only Metchnikoff's affairs but also all of Saada's debts, of which he had many. Understandably, there were no takers. The collection was stuck in a legal limbo.

Having tracked down Saada's inheritance, I couldn't take my mind off the hidden treasure in the vault. For a while, I felt like the obsessed narrator of Henry James's *The Aspern Papers.* I tried to devise strategies to gain access to the dozens of letters and documents unknown to science historians, but it all seemed to no avail. Even making copies of the documents was impossible. French law is draconian when it comes to protecting private property, though in this instance it was unclear exactly whose property was being protected.

At times, I wondered if it was all worth it. The hidden letters promised to shed light on Metchnikoff's faith in "family values"—his belief that one day, thanks to science, all family life would be a perfectly harmonious affair. But as much as I hoped to find his love letters, I also dreaded uncovering evidence of his being unfaithful to Olga. By then, I had come to feel a special bond with her in the process of delving deeply into her and her husband's lives. And I owed her a huge debt of gratitude for her biography of Metchnikoff and for having collected the material now in the Archive of the Russian Academy of Sciences, without which the work on my own book would have been unthinkable.

Then I had a stroke of luck. As part of my searches for people who had known Saada, I had found a friend of his, the caring Parisian lawyer Joseph Haddad, who had helped Saada out during his poverty-stricken years. The charming and savvy Maître Haddad took on the case as a personal challenge. But neither my biographical investigation nor the Pasteur Institute's interest in Metchnikoff's documents could supply a legal argument for opening the safes. Months went by. I worried that Metchnikoff's papers might vanish just as Aspern's had, should the bank decide to empty out the safes for which it was no longer receiving rent.

Finally, one day, Maître Haddad told me he had found the necessary loophole: clause 784 of the French Civil Code, which allowed him to deliver a triple coup—a request to open the safes so that an inventory of their contents could be made for a woman Saada had designated as his heir. At the same time, photocopies of the Metchnikoff papers could be made for my own use and that of the Pasteur Institute. A short while later, I received from Haddad the e-mail for which I'd been yearning. In a virtually unprecedented decision, the president of a regional Tribunal de Grande Instance, a French civil rights court, had granted our request.

On an overcast morning in November 2012, five years after having received Elie Wollman's clue, I entered the Crédit Lyonnais branch on the Champs-Elysées, in the rounded corner building adorned with bas-relief heads of Greek gods. I kept feeling as if I were in a movie. For one thing, I was on Paris's most iconic tree-lined thoroughfare, within a few minutes' walk from the Arc de Triomphe. In a fittingly Hollywoodish ending to the Saada saga, the safes had to be broken into. For security reasons, it normally takes two keys to unlock them—one belonging to the bank, the other to the client—but obviously Saada's keys were lost. Maître Carole Duparc-Crussard, the unexpectedly chic bailiff charged with overseeing the break-in, provided yet another cinematic touch: she arrived on a motorcycle, wearing a shiny red helmet, high heels, a miniskirted flannel suit, and a long black coat with a fox collar.

The vault itself lacked the massive fortified door of a classic heist scene. Rather, it was like a gym locker room, only with softer lighting and no benches. In the middle stood black metal tables, fastened to the floor, on which customers could lay out their valuables.

I was sure I would be ecstatic at the sight of the breached safes. Instead, after the locksmith had matter-of-factly drilled through the four locks, I was bewildered by the disarray of envelopes, shoeboxes, and fraying leather bags that turned out to be filled with the Saada family's personal archives. I had only a few hours to sift through these mounds of paper, limited by the presence of the stylish Duparc-Crussard, who was busy making an inventory of the safes' contents. In the "office" set up in the locker-lined space by the amazingly competent Pasteur Institute archivist Daniel Demellier, I frantically scanned and photographed everything that seemed historically relevant.

As I left the bank, my bounty (which, by prior agreement with the Pasteur Institute, was to remain in my exclusive possession until the publication of this book) consisted of digital copies of several hundred documents, among them more than 150 letters Metchnikoff wrote to the Rémys, which would enable me to tell the story of his love affair, as well as letters written by Olga and Lili. I quelled my uneasiness at having infringed upon their privacy with a statement Metchnikoff himself had made in one of his books: "Biographies of great people should not cover up the facts of their family life."

In addition to the digital files, I carried from the bank the thrill of having held in my hands Metchnikoff's personal belongings: a syringe; a green tin box with pills of lactic acid bacteria, forerunners of today's probiotics; and a round glass box with a sample of his hair that he had cut for his studies of aging.

In one of the safe deposit boxes, I found a gold-rimmed pince-nez I assumed had belonged to Metchnikoff. I tried them on, wondering if they would help me see today's world through his eyes—the world of the early twenty-first century, in which, after a hundred years of obscurity, his star had suddenly risen again.

II

THE MESSINA "EPIPHANY"

3

EUREKA!

In October 1882, Odessa zoologist Metchnikoff embarked on a life-altering research trip to Italy, arriving in the busy port of Messina at the foot of rocky Sicilian hills. Not yet as disheveled as in his later descriptions, his long hair combed straight back, away from his thin-rimmed spectacles, he was accompanied by his wife Olga and a brood of five of her siblings, a becoming entourage for a man used to taking his loved ones and others under his wing.

Upon sailing into the harbor, they saw a dirty quay encumbered with wooden boxes of oranges and other wares. The buildings lining the quay were also neglected and disharmonious. "Overall, Messina hardly stands out in terms of scenery, but its surroundings are highly picturesque," Metchnikoff was to recall years later in an essay about Messina in the newspaper *Russkie Vedomosti.* He and his retinue proceeded to these surroundings by carriage, traveling north along the coast to the suburban community of Ringo outside the ancient walls of Messina. They rented a small seaside cottage with a panoramic view of the bright blue Messina Strait, confined on the other side by the verdant slopes of Calabria at the southernmost tip of continental Italy. The mirrorlike flatness of the sea offered just the respite Metchnikoff needed. And he had only to cross the quay to find fishermen who could supply him with marine organisms for study.

Actually, he would have preferred to be in Naples, where in his early youth he had ventured at dawn on zoological excursions into the sea. But a cholera epidemic was raging in Naples with such force that the entire region was off-limits. Messina was an excellent second best, a prime zoological destination he had visited on a number of occasions. The funnel-shaped strait was unusually rich in marine animals.

The family had been in Messina for about three months when one winter day, Olga took her younger siblings to see apes performing in a circus. Metchnikoff was home alone in his cottage by the sea. Perhaps this sudden quiet allowed him to plunge deeper into his thoughts. After all, so many discoverers before and after him reportedly had their greatest insights while taking a peaceful break from their usual routine. (Incidentally, it was also in Sicily, in the town of Syracuse, that Archimedes is supposed to have had his archetypal "eureka moment.")

Surrounded by arrays of glassware filled with plankton-green seawater, Metchnikoff sat at a large desk with a microscope in his living room "laboratory." Holding his thumb over the top of a thin glass tube submerged into one of the flasks, he vacuum-sucked a starfish larva, *Bipinnaria asterigera,* a flat, elongated speck with several pairs of dangling extensions and a mouthlike opening on its belly. Placing it under the microscope lens, he then manipulated the tube to inject it with a drop of water containing a few grains of carmine powder. In the jellylike innards of the transparent larva, he could now conveniently observe mobile cells that had taken up the carmine, turning deep red. Metchnikoff had a special interest in these wandering cells; in worms, medusas, sponges, and other spineless animals, he had seen them gobble up food and various other particles, processing everything that got inside them.

Suddenly, he had a startling idea. "It struck me that similar cells must be serving the organism in its resistance against harmful agents," he wrote in his Messina essay. In other words, he imagined he was watching no mere act of feeding but a rudimentary form of self-defense.

"Sensing that my hunch concealed something particularly interesting, I became so excited that I began striding up and down the room and even went to the seashore to collect my thoughts," Metchnikoff wrote. "If my suggestion is true, I told myself, then a splinter in the body of the starfish larva must become quickly surrounded by encroaching mobile cells, as happens when a human being has a splinter in his finger." The larva has no vasculature or nervous system, he reasoned; if its wandering cells were to gang up on an intruding object such as a splinter, it would mean that the larva's defenses were none other than an extension of a more basic and ancient function: its primitive digestion.

"No sooner said than done. In the tiny garden of our house, where several days earlier a tangerine tree had been decorated as a 'Christmas tree' for the children, I picked several rose thorns and immediately inserted them under the skin of magnificent starfish larvae, clear like water." That "Christmas tree" makes it possible to date his experiment—arguably among the most striking in the history of science—to one of the last days of 1882 or the very beginning of 1883.

"Naturally, I was agitated throughout the night, awaiting results," Metchnikoff wrote. "The following day, in the early morning, I happily observed that the experiment had been a success." A thrilling sight greeted his eyes through the microscope lens. Just as he had predicted, masses of mobile cells inside the larvae had gathered around the intruding thorns.

There was no doubt in his mind that the cells had rushed to the larva's defense. In the same manner, they might rush to swallow another intruder, a disease-causing microbe.

The next step was almost too easy; he had no qualms about extrapolating from starfish to humans. After all, for more than twenty of his thirty-seven years, Metchnikoff had dealt with evolution of species in one way or another. He had long ago reached the conclusion that Darwin had been right. From headless mollusks to brainy mammals, all life had evolved from a common ancestor.

His daring new hypothesis was this: in all living beings, humans included, wandering cells eat up microbes, giving the organism immunity against life-threatening disease. It is these cells that are responsible for an organism's healing power. Metchnikoff had managed to observe and define a curative force, which physicians had struggled to uncover since antiquity.

The first modern theory of immunity was born.

"Until then a zoologist, I suddenly became a pathologist," Metchnikoff wrote in the final paragraph of the Messina essay. This essay, reprinted after his death in *Stranitsy vospominanii* (*Pages of Memoirs*), a collection of his autobiographical writings, is quoted in endless books and articles dealing with immunity. Paul de Kruif, in his classic *Microbe Hunters*, pokes fun at the supposed ease with which Metchnikoff interpreted his rose-thorn experiment. "It was like that blinding light that bowled Paul over on his way to Damascus—in one moment, in the most fantastical, you would say impossible flash of a second, Metchnikoff changed his whole career. . . . Nothing more was necessary (such a jumper at conclusions was he) to stamp into his brain the fixed idea that he now had the explanation of all immunity to disease."

Indeed, Metchnikoff's Messina epiphany is a good story—an idea strikes a researcher like a bolt of a lightning, instantly illuminating a new landscape of knowledge. But is it true?

In reality, Metchnikoff's eureka moment was not a moment at all. He wrote his nostalgic Messina essay in late December 1908, weeks after winning the Nobel Prize, which had suddenly turned his immunity research into front-page news. So, for the benefit of a wide audience, he resorted to a common storytelling device: collapsing all his preceding searches into a single instant of triumph. In truth, his searches had progressed more like a slow fuse, which had smoldered for years before detonating—instantaneously, indeed—in a flash of insight.

In a way, Metchnikoff had been preparing for his discovery almost from the start of his scientific career.

4

A BOY IN A HURRY

Ilya Ilyich Metchnikoff was born on May 15, 1845, in Malorossiya, or "Little Russia," part of the Russian Empire, in a region that is now eastern Ukraine. He spent his childhood in Panassovka, a small village tucked away on the undulating steppes to the east of Kharkov.

At age eight little Ilya already imagined himself a scholar. The breathtaking expanses of the steppe stretched around him all the way to the horizon as he ran down the hill from his parents' house to the nearby pond, where vegetation was unusually lush, to collect samples for his herbarium. Maples, willows, and oaks beside the pond formed magic entanglements with an undergrowth of vines and elder shrubs, perfect for exploration by an impressionable child. Nicknamed "Quicksilver," he was restless and impulsive, with silky light-chestnut hair, a rosy complexion, and sparkling gray-blue eyes. Back at the colonnaded porch of the house, he stood frowning at his "treatise" on botany, surrounded by two of his brothers and a bunch of local boys. He had paid them two kopecks each to listen to his lecture.

Metchnikoff the boy, the youngest of five siblings, was already a discoverer and a teacher—as he would be all his life—and as his lecturing to his older brothers shows, he already had a great deal of nerve.

His parents had moved to "Little Russia" from St. Petersburg, unwittingly following a motif often repeated in Russian novels: they were seeking to avoid financial ruin after his father, Ilya senior, an officer in the elite Imperial Guard, had gambled away his wife's substantial dowry at champagne-doused dinner parties. Ilya senior's estate consisted mainly of grassy pastures for horses and sheep and hardly brought any cash, but it could comfortably sustain a family of *dvoriane*, the equivalent of nobility in imperial Russia. It came with dozens of serfs, their emancipation still a decade and a half away.

As was fashionable among the *dvoriane*, the Metchnikoff clan traced its origins to a foreigner: learned seventeenth-century Moldavian adventurer Nicholas Milescu, adviser to Peter the Great, who had carried the title of Great Spatar, or "sword bearer." The name Metchnikoff, taken on by subsequent generations of the family, is derived from *mech*, the Russian word for "sword." One could hardly think of a better-suited surname for a scientist who would spend so much time sparring with his peers.

Metchnikoff's mother, Emilia, was the daughter of the Polish-born Leiba Nevakhovich, a Jewish author and businessman who in the late eighteenth century had tried for a better life in St. Petersburg, where he converted to Christianity. As generally happened to baptized Jews in Russia, he was then treated as if he had been miraculously cured of a disease; he gained full civil rights and the title of a *dvorianin*, which he passed on to Emilia and his other children. Nevakhovich's wife, Catharine Michelson, in all likelihood, was also a converted Jew.

Ilya was exceptionally close to his mother; he adored her and believed he had inherited her lively disposition. His contacts with his epicurean father, whose two greatest passions were food and cards, amounted to kissing his hand in the morning and at night.

In 1856, when Ilya was sent to high school in Kharkov—on its way to becoming one of Russia's first industrial centers—the Crimean War had just ended. The war had been a wake-up call for Russia. For the first time, masses of Russians clamored for change. Stung by their shameful defeat in the war, in which their country had tried but failed to annex portions of the dying Ottoman Empire, they demanded curbs to the tsar's arbitrary rule. Rebellious intellectuals turned to subversive Western ideas about civil liberties, social justice, and the power of science.

Much of the Western world at the time was excited about science. Steam engines pulled trains along new rails that increasingly crisscrossed thousands of miles of Europe and America; new research advances promised to optimize their performance. Cities swelled by the Industrial Revolution were counting on agricultural science to ensure their food supply through increased crop yields. In biology, German scientists had recently established cells as the fundamental units of life, heralding an unprecedented understanding of the living world.

In Russia, beyond the excitement, there was a desperate quality to the scientific fervor. In this vast agrarian country, half of the peasants still had no plows and tilled the land with more primitive tools, such as a spiked *sokha*. More than three-quarters of the population was illiterate, and nearly two-thirds of the children died before age fifteen. In these circumstances, science was perceived as a deliverance. Young men, and later women, enrolled in natural science faculties in droves in order to "serve" the people. Their desire stemmed from the messianic idea—then predominant among the Russian intelligentsia—that the educated classes, owing their prosperity to the poor masses, had to redeem their "debt to the people." And this, in turn, could best be achieved through science, medicine, or teaching. In Metchnikoff's own words, he and many others in his generation "turned to science with enthusiasm, believing that the salvation of Russia lay in that direction."

One social critic wrote that Russia seemed "to have woken from a lethargic sleep." Russian radicals wanted the old regime to go the way of last year's snow; some even dared to utter the word *republic*, circulating clandestine publications to bypass censorship.

Ilya devoured forbidden literature by flickering candlelight in Kharkov's Boardinghouse for Well-Mannered Children, where he and his brother lodged. It was with a sense of revelation that he read Ludwig Büchner's *Kraft und Stoff* (*Force and Matter*) and other books by European philosophers of scientific materialism; they believed that physical reality was all that truly existed and that science provided the absolute access to the truth. Their works were banned in Russia because they undermined religious faith—the tsar was supposed to be God's representative on earth. Needless to say, the ban guaranteed them cult status, particularly among young Russians. Ilya stopped saying prayers before going to sleep and tried so persistently to convert his friends to atheism that they gave him a nickname: "*Boga-net*," "God-is-not."

Yet people can be religious without believing in God. With a devotion that would make him the envy of any religious fanatic, Ilya turned to science for answers. He secretly read science books during such "useless" classes as theology and eliminated all extraneous activities—his only recreation was an occasional concert or opera performance. Where do we come from? Where are we headed? Above all, what is the purpose of life? These searches tortured him; he described them as a need to "answer the major questions tormenting mankind."

As happens to seekers who stumble upon the truth, at age fifteen Ilya came upon a crucial find. He saw illustrations that made him gasp: spectacular drawings of amoebae and other microscopic creatures in *Classes and Orders of the Animal Kingdom*, a hefty volume translated from German. It was only three decades earlier that microscopes had been perfected sufficiently to reveal cellular structure. Now a hidden world of tiny organisms appeared ready to yield its secrets.

Ilya's fate was sealed. He might as well have undergone an initiation ceremony. From now on, he would dedicate himself to the study of life in its most primitive forms. They could reveal truths that in humans are concealed too deeply, he reasoned—an idea that today lies at the very basis of biomedical science but was then still new.

The year Ilya turned sixteen, 1861, was a critical turning point for Russia. Two years before Abraham Lincoln issued the Emancipation Proclamation, Tsar Alexander II passed the most famous of his Great Reforms, the Emancipation Manifesto, freeing the serfs. In something like a nineteenth-century perestroika, the country shifted toward a degree of freedom, but it also seethed with unrest. Russian peasants received so little land that many believed the manifesto had been a forgery, the tsar's "real" decree having been stolen by the landlords; a few even picked up their axes to loot the estates. Radical intellectuals, equally dissatisfied, picked up their pens to demand faster reforms. Students took to the streets to demonstrate.

Like his closest friends, Ilya, at least to some extent, subscribed to nihilism, an intrepid movement that claimed part of the rebellious generation known in Russia as "people of the sixties"—the *eighteen* sixties. Perpetually angry, young Russian nihilists—children of *dvoriane,* merchants, civil servants, even clergy—negated all forms of authority: God, tsarism, the rule of landlords over peasants, of men over women, of parents over children. Science, in their mind, was the only creative force of progress, its scope extending well beyond the laboratory. It was "a weapon that can be used to distribute the fruits of knowledge, goodness, and material wealth justly among all people," the liberal magazine *Iskra* stated in 1862.

Revising everything from government to garments, the more militant nihilists even dressed to defy convention, a century ahead of "our" hippies. The males opted for freestyle artists' clothes or muzhik outfits. The female *nigilistki* replaced ruffled muslin frocks with plain dark woolen dresses. And hair? The nihilists, of course, resorted to the familiar tactic of infuriating elders with their hairdos. The men prided themselves on long, unkempt locks, and the women chopped theirs scandalously short.

Ilya was not among the extremists. One of his earliest surviving photographs shows a slender youth with delicate features, a large un-Slavic nose (in letters to his wife, Metchnikoff would self-consciously complain how prominent his nose looked in photographs), and a wistful gaze romantically directed into the unknown. His fine, wavy hair is long but neatly parted on the side. Obviously posed, he wears his square Sunday finest—black jacket, starched white shirt with stand-up collar, and bow tie. It's the kind of photograph he could send to a fiancée, if he had one, from a respectable family like his own.

But even though there was no hint at rebellion in his appearance, it is clear that Ilya was a nihilist based on an essay he wrote later in life about the history of science in Russia. The figure that embodied the spirit of his generation, in his mind, was the "nihilist" Yevgeny Bazarov—hero of Ivan Turgenev's influential contemporary novel *Fathers and Sons*—after whom the term *nihilists* was applied to Russia's mid-nineteenth-century rebels. Bazarov was a gruff young doctor who believed only in science. In his essay, Metchnikoff makes an announcement that would have surely made Bazarov proud: "There's no doubt whatsoever that in science, as in practical life, one should never bow to authority of any sort." Nor is there any doubt that nihilism helped him mature into a decidedly antidogmatic scientist.

Already in high school, Ilya acted as if time was running out. A sensitive boy, quick to be hurt and just as quick to lose his temper, he sensed the scale of his own talent very early on. Lacking the patience to wait until adulthood, he was in a hurry to save Russia and humanity—preferably by making a great scientific discovery. Shedding his school uniform to change into regular clothes so that he could pass for a university student, he sneaked into lectures at the University of Kharkov. Borrowing a microscope, he began to study single-celled aquatic creatures. And he wanted to make his opinions known: at sixteen, he published his first article, a review of a geology textbook.

After finishing high school at age seventeen with a gold medal, the highest honor, he intended to study medicine, which was then the top choice for young Russians seeking to improve the lot of humanity. He felt inspired by the cell theory of disease, recently created by the celebrated German pathologist Rudolf Virchow. By showing that cells were altered in disease, and that by examining cells physicians could diagnose cancer and other disorders, Virchow had turned the entire medical world into his disciples. Ilya dreamed of creating his own theory in medicine. He wanted to have an impact on medical science on the same scale as Virchow.

But his mother urged him to aim for a career in research, not medical practice. "You are too sensitive to view the constant sight of human suffering," she said. Perhaps. Within the next two years, Ilya whisked through a four-year course in the natural sciences at the University of Kharkov.

5

SCIENCE AND MARRIAGE

AT TWENTY, AN AGE WHEN most young scientists are still mastering the basics, Ilya embarked upon research that would soon turn him into a founder of a new branch of science.

Shortly after graduation in 1864 he went abroad on a brief research trip, but he then managed to obtain a scholarship through an enlightened, if short-lived, program the tsar had created at the height of Russia's liberal spring. Several dozen young Russian scientists were sent to study abroad so that upon return they would help revamp the empire's universities. Ilya's trip extended to three years of study and wanderlust and became his own version of a grand tour of Europe. Its highlights were not the Alps or the ruins of Pompeii but some of the best German university laboratories and, on the islands of Heligoland and in Italy, several beaches most suitable for studying invertebrate creatures.

The bay of Naples, then Italy's largest metropolis, was so rich in marine life that it was a favorite destination for *dottori dei pesci*, or "fish doctors," as locals called visiting zoologists. Ilya's thin, energetic figure, five feet nine inches tall, dragging around a bucket of seawater or, like a butterfly hunter, carrying a net on a long pole, became a fixture on the quays of Naples. At four in the morning, he hired a fisherman to catch him inedible *frutti di mare*—sponges, sea squirts, comb jellies, medusas. Back in his rented room, as he dissected them and their embryos, he posed the grand questions. Does nature have a hidden blueprint? Can the riddle of life be teased out of life's simplest forms?

Like many zoologists of his day, Ilya searched for the secret order in the animal kingdom. How are living creatures related to one another? Had they all, including humans, evolved from a common ancestor as Charles Darwin had suggested? Darwin's *On the Origin of Species,* published more than five years earlier, had left him spellbound. In Metchnikoff's own words, "Darwin had fostered the hope for solving the problem of human existence through studying the laws governing living beings." But Darwin's belief in the common ancestor was still hotly disputed. Some of the leading scientists thought different types of animals had their own separate ancestries. In Naples, Ilya settled on a deserving area of study that he hoped could help resolve the controversy: embryos of invertebrates. For his master's thesis he dissected the larvae of the dwarf bobtail squid.

But upon Ilya's return to Russia at age twenty-two, his high hopes began to crash against reality.

The elongated red-and-white baroque building of the Imperial St. Petersburg University stood, as it does today, on a quay overlooking a breathtaking skyline of domes, spires, and monuments across the Neva River. On the quay itself, two granite Egyptian sphinxes, purchased in Alexandria on the tsar's orders, stare in scorn at the icy river. In 1868 Ilya obtained the post of *dotsent* in zoology at the university, something like an assistant professor. Inside its famously long hallway, he could run into quite a few upcoming stars of Russian science, among them the young chemistry professor Dmitry Mendeleyev, who had just invented a clever teaching device for beginning students: a periodic system for classifying elements by atomic weight.

Ilya trudged through St. Petersburg's poorly lit cobbled streets in bone-chilling cold, whistling his favorite aria from *The Magic Flute.* "Were I as small as a snail, I would creep into my shell." (A passionate lover of music, especially opera, he could whistle from memory any musical piece he had ever heard.) The exuberant young man, who had only recently imagined himself engaging in the "salvation of Russia," was bitterly disappointed. To begin with, where were the spacious, fully equipped laboratories he had encountered in Germany? His only workspace at the university was a desk squeezed between cabinets in the unheated zoological museum. His salary was meager. He expected older scientists to nurture his talents, but encountering indifference, he grew bitter and suspicious, imagining that they were plotting against him. Then there were his eyes. Frequently inflamed for a reason that never became clear, they hurt terribly. Would he have to give up science? This horrifying thought tortured him during sleepless nights.

He found solace in longing for love and marriage. Some of his closest friends were then entering into fictitious marriages. This was a strategy adopted by Russian nihilists for "liberating" young women. A married *nigilistka* gained freedoms denied to single women—she could move to the city, enroll in a university, or travel abroad without disgracing her family. In Nikolai Chernyshevsky's 1863 utopian novel, *What Is to Be Done?*, which was a gospel of Russian radicals, the heroine indeed approached marriage as a means to pursue her socialist ideals. Overall, the nihilists viewed marriage as a form of oppression, ultimately to be abolished. But Ilya, surprisingly conservative in his intimate yearnings, wouldn't even consider the short-haired *nigilistki* as potential mates, fictitious or not. His utterly unnihilist desire was to find a real wife.

Ilya had a Pygmalion-like fantasy of molding a young girl, the daughter of a botany professor at the university, in his own erudite image and marrying her one day. But a quiet young woman his age, the professor's niece Ludmila Fedorovich, entered the scene. She cared for him when he was ill and cheered him up when he poured his troubles out to her, which was often. In turn, when Ludmila became ill, Ilya cared for her. Their friendship soon blossomed into love. Alas, their 1869 marriage only hastened his descent into despair.

Their union was off to a sinister start. The bride, in a lilac silk dress, had to be carried to the church in a chair. At first she was diagnosed with flu, but it then turned out that she suffered from a ubiquitous nineteenth-century scourge, consumption, as tuberculosis was then called. Romantic tubercular heroines commonly figured in novels and on stage. One was Violetta of Giuseppe Verdi's *La Traviata*, which Ilya might have seen in the theater, as it had just then been produced in St. Petersburg for the first time. But when he hurried home to his own

Violetta, there was nothing romantic about finding his frail young wife wrapped in a shawl, shivering and spitting blood.

None of Ilya's devoted efforts—neither moving to the warmer Odessa, where he became professor at Novorossiya University (after obtaining a doctorate for his work on small crustaceans), nor spending all his salary to send his wife for "climate therapy" abroad—helped to save Ludmila. "A grave covered with flowers," Ilya thought about the Portuguese island of Madeira, famed for its mild climate, where he took her while still clinging to a last hope. In April 1873, she died there.

For a while, Ilya numbed his pain with Ludmila's morphine. Stopping in Geneva on his way back to Russia, he felt exhausted, physically and mentally. Ludmila's cruel death had robbed his life of purpose. His eyes were so inflamed he was occasionally forced to wear a bandage that made it impossible to read or write. His career had halted precariously; he had recently been passed over for a prestigious prize that he had won in the past.

"Why live?" he asked himself. "My personal life is finished. My eyes are so poor I'll soon go blind."

Metchnikoff described his next desperate step as "my Geneva catastrophe." He swallowed his entire supply of morphine in an attempt to kill himself. But the dose was so large it caused vomiting. His body purged the poison.

He tried another tack at killing himself: after a scalding hot bath, he took a freezing cold shower and went out into the chilly night air, hoping to catch a lethal cold. Instead, on a bridge over the Rhône River, he suddenly saw a swarm of winged insects around a lantern. His thoughts switched from the meaninglessness of his own life to the brevity of the life of those creatures, which hardly left the time for natural selection to affect their evolution. Many years later, Olga summed up the end of the crisis in her book, *Life of Elie Metchnikoff*: "His thoughts turned to scientific issues. He was saved. A connection with life was reestablished!"

Back in Odessa, a widower at twenty-eight, Ilya felt so fragile he avoided going to tragedies in the theater, lest he burst into tears. But he pressed on with research that on three occasions earned him the triannual Karl von Baer Prize of the Imperial Saint Petersburg Academy of Sciences, as well as international fame. He had been drawn to his central topic, embryology, by an idea that promised to solve the ancestry issue of different groups of animals. The idea, proposed by a German zoologist and known as the recapitulation theory, was that during embryonic development, each animal goes through the stages that recapitulate, in compressed form, the entire evolutionary history of its species. Only a most limited version of the theory was eventually proved to be true (many embryos pass through stages that correspond to the embryonic development, not to the adult forms, of their ancestors), but for a while it captured many minds, including Ilya's.

Scientists had already demystified the embryonic development of fish, birds, humans, and other vertebrates some three decades earlier, showing that it obeyed strict laws—all the embryos grew from the same three germ layers, each containing a germ of distinct organs. But knowledge about the embryos of the vastly more numerous invertebrates was still fragmentary and focused mainly on appearance, not development, so that comparing different species was impossible. Ilya was able to make such comparisons by tracing different stages of animal development. That was how, with studies he began on his grand tour and continued at home, he became a cofounder of a new area of research: comparative embryology of invertebrates.

In dozens of brilliant studies spanning some two decades, Ilya ultimately unveiled the embryology of invertebrates in over two dozen families, tracing the development from egg to adult in worms, sea urchins, brittle stars, gelatinous medusas, and luminescent bell-shaped siphonophores. In head-footed cephalopod mollusks scooped out of the warm Mediterranean waters, he found the first definitive evidence that embryos of invertebrates had germ layers, just like those of vertebrates—a striking proof of their common ancestry. He then found germ layers in the embryos of such disparate animals as scorpions and crustaceans, becoming a convinced Darwinian in the process. His research on marine animals called *echinoderms,* in which he established their evolutionary link to comb jellies and medusas, was to be labeled "epoch-making" by embryologists.

From the very start, uncommon combativeness marked his scientific career. Already in his first serious scientific paper, published at age eighteen as a freshman in the natural sciences at the University of Kharkov, he had challenged the findings of a famous Berlin physiologist on the nature of the stalk in the microscopic freshwater creature *Vorticella*. His pugnaciousness continued to erupt throughout his grand tour. Whenever he published research results, he disputed the opinions of others. When a prominent German scientist failed to acknowledge his contribution to a study, Ilya produced an angry brochure to claim proper credit for his work.

Writing to Ilya from St. Petersburg, eminent naturalist Karl von Baer, a founding father of the entire field of embryology, implored the younger man, "I delight in your energy and hope you do honor to your country. But would you pay heed to an old man, who has already totally overcome his own ambition, and respect his wish that you cut down on polemics with others." Ilya paid no heed.

Even when not actively arguing, he was short-tempered. He threw a tantrum in one laboratory in Germany when his clumsy hands refused to obey him as he tried to dissect a rare lizard. He tossed the precious reptile, along with the cutting tools, onto the floor. On another occasion, in Italy, when incessant serenades below his window interfered with his sleep, he emptied a chamber pot of liquid waste onto the heads of the hapless musicians. In Odessa, when the scientist husband of one of his relatives claimed to have achieved spontaneous generation of bacteria in distilled water, an irate Ilya, trying to convince the scientist that the germs had "generated" from his dirty fingers, hurled a laboratory vessel at him.

Later in his career, Ilya's combativeness was to prove both a hindrance and a blessing.

6

A PERSON OF EXTREME CONVICTIONS

ONCE DURING A COUNCIL MEETING at Novorossiya University, Ilya jumped on one of the room's carved-legged wooden tables in the heat of an argument to get everyone's attention. His contemporaries in Odessa cited this episode as evidence of his arrogance and the difficulty in dealing with him. This reputation didn't prevent him from being the pride of Odessa. His lectures to the general public—he ardently believed in spreading the word about science—drew enormous crowds, rain or shine.

Odessa was then a welcoming, cosmopolitan city, lovingly called "Odessa-mama" by its patriots. Greeks, Italians, other foreigners and Jews were permitted to settle here by the tsar as part of his efforts to populate southern Novorossiya, or "New Russia." They exported grain, imported wine, and built roads and railroads. Odessa's population had been doubling every two decades, so there was a "new" look about its straight, broad streets. A Russian geographer who visited Odessa in the late 1860s wrote that by then it "had turned into quite a European city, with stone, even granite carriageways, large omnibuses, good, if pricey, hotels, rapid steamboat lines, and other amenities of a big population center."

The Novorossiya University, which today bears Metchnikoff's name, occupies a yellow-and-white cream cake–like mansion in

Odessa's historic heart. True to his sense of mission and the desire to build up Russia's science, Ilya worked hard to turn it into a center of research rather than only of teaching. This task required recruiting worthwhile researchers in all disciplines, offering Ilya plenty of opportunities to apply his combative spirit in dealings with the university's administration.

But when Ilya was surrounded by his intimates, you would have never known he was considered difficult. Whereas with strangers he was chronically on guard, ready to launch a preemptive strike, his prickliness vanished among close friends. Warm, talkative, and funny, he obsessively cared for all the nearest and dearest who were, or weren't, in need of his help, worrying about their health and tending to their daily needs. The Russian anarchist Mikhail Bakunin, whom Ilya had befriended in Italy while on his grand tour, had even called him "*Mamasha*," or "Mommy," a nickname later picked up by some of his Odessa friends.

Ludmila had been dead for nearly two years when Ilya called upon his upstairs neighbors, wealthy *dvoriane* landowners called the Belokopytoffs, to beg for quiet; their eight children were generating considerable racket. But one of their twin oldest daughters, Olga, a tall, pretty blonde who had just turned sixteen, was to disturb his peace in more ways than one. Olga, for her part, was taken with the pale, thin neighbor. "I thought he looked rather like Christ," she wrote in her book. Their marriage in February 1875 raised eyebrows. Nihilists believed in romantic unions between equals, whereas in marrying Olga, nearly fourteen years his junior, Ilya was finally acting out his Pygmalion scheme, even teaching his young wife biology.

With her long, braided locks and her quiet resolve, Olga typified a particular kind of Russian feminine ideal still known today as a "Turgenev maiden." Not too prone to having fun, these fair-haired young women are pure, modest, and devoted to higher goals. Olga's ultimate, self-sacrificial devotion was to Ilya. The day after the wedding, she gave her new husband a surprise, and quite a surprise it was. She did not cook a delicious meal but served him something he appreciated much more: fish bones, demonstrating her savvy about the development of fish vertebrae. "Despite all the excitement of the wedding, the following morning I got up at dawn to prepare my zoology lesson," she recalls in the biography she published after Metchnikoff's death.

The four hundred or so letters Metchnikoff wrote to Olga in over forty years of marriage read like a diary, centered mainly on his work. "My sweet girl, my dove, my gold, my beloved baby," opens a typical letter. It is not, however, in these expansive terms of endearment, but rather in their main text that the yellowed sheets of paper, inked in Metchnikoff's lacelike handwriting, reveal the essence of his intimacy with Olga: his ability to share with her his obsession with science. There was no doubt in his mind that Olga's greatest joy was to take pride in his work. "I was thinking about my dear one, and what I could do to please her most," he wrote to Olga shortly after their wedding, referring to her in the third person. "I've decided the best thing would be to prepare my public lecture as soon and as diligently as possible."

What helped Olga to keep her focus on Ilya and his work was that they had no children. "Life shouldn't be valued, creating new living beings is criminal," Ilya had announced in his gloomiest days. Even if Olga tried to protest, she was up against a dual obstruction: doctors had advised her against giving birth. (The reasons for the injunction are unknown, but it was probably justified: Olga's identical twin was to die in childbirth.)

In the spring of 1881 two nearly simultaneous events, one private, the other public, changed the course of Metchnikoff's life.

The private event was one of the most enigmatic of all the fits and starts that characterized his entire life. In the previous year, when Olga had contracted typhus, he overexerted himself in care and worry. For months after she recovered, he suffered from heart pain, headaches, dizziness, and insomnia. One day, when he thought his speech was slightly slurred, he feared it was the beginning of a partial paralysis. "In the throngs of nervous angst, he decided to commit suicide," Olga wrote in her book, the main source for the bizarre episode that followed. Thirty-five-year-old Metchnikoff, who already had a history of failed suicide, chose a method that was by no means a sure way to die: he injected himself with the blood of a patient sick with relapsing fever, then a mysterious plague. (It is now known to be a bacterial infection spread by lice or ticks.) The ostensible goal was to resolve the question of whether this disease was transmitted by blood. Did he indeed mean to kill himself? Or was this a reckless case of self-experimentation, a Russian roulette into which, in a moment of nervous exhaustion, he threw himself with suicidal abandon?

In any event, for a future founder of immunology, his startling choice proved a fitting test of his body's own immunity. Its consequences bring to mind ancient rites in which shamans, whose job in society was to heal others, inoculated themselves to enhance their own special gift of healing. Ilya did not believe in magic, but his faith in science seemed to have supplied him with an equivalent of such a ritual. He came out of his self-experiment with renewed powers.

At first he became gravely ill, but the self-inflicted illness triggered an acute eye inflammation that not only failed to damage his eyes but actually cured them. He could get away with staring into the microscope for hours for the rest of his life. No less important, when his strength came back he felt a thirst for life he had never known before. A near brush with death often alters people's perspectives, but for one as sensitive as Ilya, it proved a life-changing experience. "After recovering from relapsing fever, he went through a sort of a general rebirth: he developed a tremendous love of life, his health flourished, his energy and work capacity were greater than ever, the pessimism of his youth finally started to wane, paling before the nascent optimistic dawn of his mature age," Olga wrote. It was the kernel of his optimistic worldview that some two decades later was to become his credo.

When Ilya still lay delirious in bed with a 104-degree fever, he received thunderbolt news: members of the terrorist organization Narodnaya Volya, "People's Will," had thrown two bombs at the tsar's carriage in St. Petersburg, assassinating Alexander II. They had been hoping to start a revolution. Instead, they ushered in a period of intense oppression. The brutal Alexander III, who believed his father's Great Reforms had been a bad idea, promptly reversed the recently granted freedoms. He had members of Narodnaya Volya executed and newly arrested "disloyal" persons thrown into prisons or shipped off to Siberia. The crackdown on universities, viewed as hotbeds of sedition, was particularly thorough. Students were banned from holding all gatherings, and the minister had to approve every decision of the university council.

In Odessa, in a list of "politically untrustworthy" faculty he had compiled, the chief of the gendarmerie, one Colonel Pershin, declared, "Metchnikoff, a professor of zoology, a person of extreme convictions, impossible in any educational establishment." In reality, all his life, Metchnikoff was a moderate liberal. Abhorring violence, he was against revolutions. But in tsarist Russia, it didn't take much to be labeled a dangerous radical.

Metchnikoff, for his part, was finding that his struggles, difficult until then, had become impossible. For all his protests, Novorossiya University was hiring new faculty members based on political loyalty, not scientific merit. The university appointed two mediocre professors, one in chemistry, another in law, despite his objections. A Jewish lecturer in anatomy who refused to convert to Christianity, and whom Metchnikoff had been protecting for more than a decade, was finally fired. "Attending council meetings became real torture," he wrote in his essay "Why I Settled Abroad."

His dealings with students were also a permanent source of aggravation. Instead of fostering their—and his own—scientific aspirations, he was constantly defending them or freeing them from jail. Then a new rector created such a difficult regime that conflict became inevitable. Once Metchnikoff was summoned to intervene when a shouting crowd of students had occupied the university's upper floor, protesting against a dean's decision to reject a dissertation as "too socialist." The students stopped rioting after Metchnikoff negotiated an agreement with the district warden on their behalf, but it soon became clear the warden had no intention of keeping his promises. "The warden's betrayal was the last straw," Metchnikoff wrote in the essay.

It was late spring, and Odessa was at its prettiest. Jarring with the somber mood at the university, blooming locust trees filled its streets with a fragrance reminiscent of orange blossoms. On May 22, 1882, Metchnikoff submitted his letter of resignation. Shortly afterward, so did at least three other liberal professors. "The mere memory of Novorossiya University causes me to shudder and makes me ill," Metchnikoff wrote to a friend several months later.

He was never to assume the post of regional entomologist that he had secured prior to his resignation: just then, both of Olga's parents died. She inherited a share of the family's agricultural estate near Kiev that brought in 8,000 rubles a year, almost three times a professor's salary. Freed from university vexations and financial worries, Metchnikoff was finally at liberty to plunge into research to his heart's content.

At thirty-seven, he had already accomplished enough to sustain a stellar career but he was far from ready to put away his microscope. And he had yet to fulfill his childhood dream: creating his own medical theory, like the one by Virchow that had once dazzled him.

7

THE TRUE STORY

WHEN IN THE FALL OF 1882 Metchnikoff traveled to Messina with the newly inherited money, he already had a research program in mind. As mentioned earlier, the discovery he made on that trip had been in preparation for nearly two decades.

At age twenty, in Germany, while on his grand tour, he had picked up a minuscule hairy worm in a greenhouse of the Giessen botanical garden. Studying it under a microscope, he was struck by the worm's peculiar way of processing food. Its body lacked a proper gut but was filled with cell-like structures that directly absorbed food particles. Until then, Metchnikoff had seen such primitive digestion only in single-celled animals. The worm was to prove a prize find, though at the time of his grand tour he had no way of knowing how great the prize.

In the following years, he continued to come across similarly gluttonous cells again and again in various creatures. He also found that in their earliest stages, the embryos of certain invertebrates resembled a sac filled with a mass of such cells, which served mainly to digest food.

He continued peacefully observing these cells until a young, furiously pro-Darwinian German scientist named Ernst Haeckel published a hypothesis about the very first multicellular creature that once upon a time had appeared on Earth, which was until then populated by single-celled organisms alone. Haeckel, nicknamed the "German Darwin," ventured a guess that this mysterious creature, which he called *gastrea*, must have resembled a microscopic, double-bottomed sac, a structure familiar from the early stage embryos of numerous species.

Metchnikoff seized this opportunity to pick one of his many scientific fights. The ancestor of multicellular life, he argued, had not been Haeckel's gastrea at all, but rather a tiny sac filled with voracious cells, a structure he had seen in embryos of animals more ancient than the ones cited by Haeckel. In lectures and in writing, Metchnikoff lashed out against the "German Darwin" with unusual spite. He stated in his "Essay on the Question of the Origin of Species" that Haeckel was a dilettante, that he had "more diligence than talent," and that his work methods were "unreliable, untimely, and unscientific." Metchnikoff even altered the direction of his own research, focusing on the voracious wandering cells in order to counteract Haeckel.

The timing of this unresolvable controversy, and of the beginning of Metchnikoff's obsession with the wandering cells, squarely supports what present-day psychologists call the ten-year rule: that even the most creative individuals must immerse themselves in a discipline for at least a decade before producing a world-class breakthrough. In fact, this rule is regarded as the only widely respected "law" of creativity in music, art, and science, Andrew Robinson wrote in his book *Sudden Genius?*, in which he quoted an anonymous Greek poet: "Before the gates of Excellence the high gods have placed sweat." Haeckel published his gastrea hypothesis in 1872—exactly ten years before Metchnikoff's Messina "epiphany."

His combative juices stirred, Metchnikoff began to seek out the most ancient invertebrates on the assumption that they should still have traces of the primitive, cellular digestion. Indeed, he found this digestion to be prevalent in lower worms, in bottom-dwelling sponges that had inhabited the seas for nearly a billion years, and in such primordial animals as medusas, comb jellies, and siphonophores. It was with the express purpose of studying wandering cells in ancient marine creatures, and with the latent purpose of beating Haeckel, that he had come to Messina in 1882.

In Messina, he soon found that these cells reveal their appetite not only in digestion but also in quite a few other vital phenomena. When a starfish larva metamorphosed into an adult animal, mobile cells ate up the surplus tissue. He later found that the same happened when a tadpole turned into a frog. But Metchnikoff knew that the cells were also capable of gobbling up foreign particles from outside the animal's own body. It was only natural that in search of additional roles for the cells, his thoughts turned to immunity against infection.

It is true that before Messina, he had not directly concerned himself with immunity. The body's ability to conquer disease was the realm of pathology, the medical specialty dealing with disease mechanisms, quite distant from his zoological pursuits. But Metchnikoff was an immensely curious man; he spent every free moment reading, staying on top of advances in many areas of science. And he was forever concerned with humanity's health and well-being.

In immunity, he finally found a topic worthy of his prodigious talent.

8

AN OUTSIDER'S ADVANTAGE

AS METCHNIKOFF HIMSELF WROTE IN his treatise *Immunity in Infective Diseases,* "protection against disease is one of the most crucial issues ever to have preoccupied humanity, so it is only natural that great attention should have been devoted to it from the most remote times." Indeed, since antiquity, kings, shamans, and physicians had searched for and sometimes succeeded in producing immunity against snake poison or various diseases of humans and cattle.

More than two millennia ago, the worst fear of Mithridates the Great, the paranoid king of Pontus, was that he would succumb to the same fate as his father, the elder royal who died in horrible convulsions after tasting a poisoned dish. Mithridates attempted to harden himself against all earthly poisons. Each day he swallowed a toxic mixture of pounded-up herbs rendered palatable by a few drops of honey. The potion reportedly worked too well for the king's own good. Facing capture by the Romans, Mithridates was unable to commit suicide by drinking poison—he had rendered himself immune.

Royal predicaments aside, the notion of immunity cropped up throughout history in connection with such devastating diseases as smallpox, "the speckled monster," which at times, by some counts, left one-fifth of the world's population dead or disfigured. The disfigurement created an advantage: people who survived a bout of smallpox did not get sick again. So as early as the tenth century CE, wise men in China learned to inoculate people against this disease by using secret techniques handed down from one practitioner to another. For instance, they blew flakes from smallpox scabs through a silver tube into the nose—into the left nostril for boys, the right one for girls.

The secret must have been hard to keep, or perhaps it kept being reinvented in different corners of the world. In any event, by the eighteenth century folk healers in parts of Asia, Africa, and rural Europe were openly performing inoculations against the dreaded smallpox. The poorer Circassians in the Caucasus Mountains and in Constantinople had a dark reason for inoculating their daughters. Fabled for their beauty, young Circassian women were greatly valued as concubines, but if pockmarked, they could not be sold off into Turkish harems. "The smallpox getting into the family, one daughter died of it, another lost an eye, a third had a great nose at her recovery, and the unhappy parents were completely ruined," Voltaire wrote to a friend in 1753 in his letter "On Inoculation." To protect such candidates from future disfigurement, six-month-old Circassian baby girls had their tender skin pricked, then smeared with discharge from smallpox scabs, the punctures covered with green leaves and patches of sheep skin. It was indeed in Constantinople that the wife of the British ambassador, Lady Wortley Montagu—whom Voltaire described as "a woman of as fine a genius, and endued with as great a strength of mind, as any of her sex in the British Kingdoms"—had her infant son inoculated against smallpox. She later started an "engrafting" craze among the British aristocracy.

As Metchnikoff wrote in his *Immunity* treatise, such risky inoculations were finally ended in the late eighteenth century by Edward Jenner, the "highly intelligent and educated" country doctor from Gloucester. Jenner learned that milkmaids who had caught the relatively benign cowpox from cattle became immune to smallpox, and thus he prepared a vaccine—a word he coined from the Latin *vacca*, "cow"—from cowpox sores. His vaccinations were ultimately adopted not only throughout the British Empire but even in France, despite an ongoing war with Britain. Napoleon himself had a medal minted in Jenner's honor.

In the nineteenth century, it was France's turn to take the lead in inoculation. Its immunization impresario Louis Pasteur so admired

Jenner that at an international medical congress in London, he suggested using the term *vaccination* for all forms of inoculation—including those that had nothing to do with cows—in honor of "the merit and immense services rendered by one of the greatest men of England, your Jenner."

In May 1881, when Metchnikoff and Russia were busy coping with the aftermath of the tsar's assassination, Pasteur stunned the world with a public experiment in the French village of Pouilly-le-Fort by demonstrating the efficacy of a new vaccine that could render farm animals immune against anthrax, then a widespread agricultural disease. Pasteur's method worked on the same principle as Jenner's: weakened germs produced immunity against the disease caused by their robust counterparts. The vaccine "merely communicates to animals a benignant malady, which preserves them from the deadly form," Pasteur wrote.

But how the vaccines rendered cattle or people immune was a mystery. As often happens in medicine, theory lagged behind practice.

Throughout history, there had been no shortage of interesting theories of immunity. In the ninth century, one medical authority in Persia explained immunity to smallpox: excess moisture is driven from the blood through the smallpox pustules on the skin; whoever survives the ordeal no longer has the moisture for an additional bout of the disease. Six centuries later, a Renaissance scholar in Italy claimed that contaminants of menstrual blood, with which we are all born, were expelled through the smallpox pustules. Once this impurity was gone, so was the substrate for further smallpox attacks. One eighteenth-century explanation of immunity, by an English physician, was strictly mechanical. The formation of the pocks in smallpox supposedly distended the pores of the skin, causing them to stay forever open and thereby preventing morbid matter from accumulating. Yet another (somewhat horticultural) theory, also from England, held that we are all born with the seeds of every possible disease, which deplete themselves once the disease has run its course.

When the germ theory—of which Pasteur was a major champion—took hold in medicine, the quest to understand immunity had entered a new phase. It turned out that contrary to ancient belief, perhaps not entirely stamped out until today, disease had nothing to do with punishment. Germs, not gods, were definitively pinned down as pervasive culprits. So Pasteur and other medical researchers began to tackle the problem of immunity, natural or acquired, from the vantage point of the germ theory. When microbes raided the body, what determined the outcome of the invasion? How did vaccinations help the body end the microbial infection? And what would it take to make us resistant to germs once and for all?

Annoyed at being unable to explain how his own vaccines worked, Pasteur ventured a hypothesis. He suggested that they helped to starve microbes to death. The vaccines, he said, contained weakened germs, which supposedly used up certain nutrients. By the time real, disease-causing germs invaded the body, they had nothing to eat. As a result, body tissue becomes "in some way incapable of supporting the growth of the microbe," Pasteur told members of the French Academy of Sciences in 1880.

But Pasteur wasn't quite happy with his own hypothesis. For starters, some vaccines were shown to work even if they contained not weakened but dead microbes, which obviously couldn't consume nutrients. Besides, his hypothesis could explain immunity resulting from vaccination or prior exposure to a disease, but why were some people immune to infection even without any vaccination or prior exposure?

Noted Lyon veterinarian Auguste Chauveau tried to shed light on such natural immunity. He proposed an explanation that contradicted Pasteur's hypothesis. Microbes did not die of starvation, he argued; rather, upon entering blood, they encountered a substance that could destroy them. Why did this natural immunity sometimes fail? When there were too many microbes, Chauveau said, the protective substance ran out. As for acquired immunity, German pathologist Paul Grawitz also offered an alternative explanation to Pasteur's. After an initial bout of illness, Grawitz suggested, people acquired immunity to infectious agents due to "adaptation of the cells to an energetic assimilation of the fungi."

If there was a unifying feature, apart from fogginess, to these and other theories of immunity from the same period, it was their view of the human body as a passive and inert receptacle. No inner mechanism was known to produce the body's resistance, and so it remained an enigma.

In contrast, scientists were making spectacular advances in revealing the causes of infectious diseases. In the spring of 1882, a few months before departing for Messina, Metchnikoff was stunned by a report from Berlin. Robert Koch, a doyen of bacteriology, had discovered the fine microscopic rods causing tuberculosis, the century's leading killer in the Western world. Unaware of how much grief Koch would cause him in the future, Metchnikoff was rapt. Besides, the discovery could have easily touched a personal note; he had lived with a tubercular wife for four years. Perhaps he began to wonder why he had not fallen ill. How had his body protected itself against the tuberculosis germs?

Whether his fascination was personal or not, Metchnikoff grew increasingly interested in human diseases. From his *Immunity*, in which he provides a much more gradual account of his Messina discovery than in the oft-quoted essay, we learn that before leaving for Messina, he had attended a course in pathology. In that course he learned that in inflammation, white blood cells, or leukocytes (from the Greek *leuko*, "white"), seep through the walls of blood vessels and accumulate at inflamed sites. And inflammation was thought to be most often caused by microbes, which, in turn, were occasionally found inside the leukocytes.

In Messina, it all began to come together. On the one hand, starfish cells ganged up on chunks of food, useless tissue, or external debris. On the other, human white blood cells moved to the site of infection. That was when the slow-burning fuse triggered an explosive revelation. Metchnikoff's intuition hatched an eerily accurate insight into the workings of the living world. He realized that certain mobile cells in vertebrates are evolutionary heirs of single-celled creatures.

On the lower rungs of life, the distinction between digestion and protection is blurred. In fact, in the amoeba, consuming food or foe is one and the same activity. The first multicellular organisms to appear on Earth must have possessed a number of cells that gobbled up food and detractors in the same manner as did single-celled organisms. In simple invertebrates that have no gut, these cells still function as a digestive system. But as life evolved, the two functions diverged. In vertebrates, including humans, that have inherited these same cells in the course of evolution, the cells continue to eat, only they now eat infectious invaders. These are specialized cells devoted entirely to protection.

Had Darwin not died a few months earlier, he would have surely been thrilled. This revelation created a uniting thread between single-celled organisms and humans. Metchnikoff's teenage conviction that life's simplest forms could shed light on humans had paid off.

Scores of researchers before him had seen the mobile cells filled with microbes. The prevailing view was that the microbes had passively drifted into the cells in the course of inflammation, which, in contrast to Metchnikoff's hypothesis, was viewed as harmful. Unbeknownst to Metchnikoff, a handful of researchers had actually suggested that the cells might be destroying microbes, but their voices were drowned in the cacophony of speculations about the nature of immunity. He was the only one to back his hunch with an entire theory of immunity—and to turn its development into his life's work.

Indeed, the vast majority of Metchnikoff's contemporaries granted mobile cells no role whatsoever in defending the body. Why did Metchnikoff see something practically no one else did? How did a zoologist propose the first modern solution to an ancient riddle that had stumped physicians over the millennia?

Paradoxically, being a zoologist rather than a physician turned out to be an advantage. Working outside his own discipline, he was unencumbered by the conventions that inevitably build up over time within any field. He thus remained free to unleash the imagination that helped him make one of the greatest discoveries in the history of medicine.

As for his character, his nihilist nerve was yet another indispensable ingredient in the making of that discovery. It helped him bring together recent radical ideas—the cell theory of disease, the germ theory, Darwin's theory of evolution—into a brave new theory of his own. And for once his preemptive nature served him well. Spoiling for combat, he evidently recognized his own acute need for self-defense in the spectacle of voracious cells. This universal tendency to perceive the world through the prism of one's own sensitivities is perhaps best summed up by the popular saying "It takes one to know one." In Metchnikoff's case, it took a fighter to know a fighting mechanism.

9

EATING CELLS

METCHNIKOFF WAS SOON TO LEARN the hard way that in science, as elsewhere, it's not enough to be right. No less crucial is to be believed. But beginner's luck smiled at him. In Messina, he encountered a deceptively warm reception for his new hypothesis.

It so happened that the very first person with whom he shared his insight about wandering cells was not one to be put off by maverick ideas. Nicolaus Kleinenberg, a native of the Baltic region, where people are reputed to be headstrong, was the goateed author of classic studies on *Hydra.* He had ended up at the University of Messina after falling out with colleagues on the Continent, and he enthusiastically welcomed Metchnikoff's bold proposal to link wandering cells with immunity.

"*Das ist ein wahrer Hippokratische Gedanke!*" (This is a truly Hippocratic thought!) he exclaimed.

Trained as a doctor, Kleinenberg was of course referring to the Greek physician's belief in the inner curative force. Hippocrates, undisputed father of Western medicine, had argued that the human body contained four basic substances called *humors*—blood, phlegm, and yellow and black bile—which had to be properly balanced. While perfecting his art of healing on the golden-beached Greek island of Kos, Hippocrates had come to admire the wisdom of nature, which he saw as the body's natural ability to rebalance the four humors. "The body's nature is the physician in disease," he wrote famously. "Well trained, readily and without instruction, nature does what is needed." His own and future generations took these pronouncements to mean that each person's body had its own curative force, as wondrous as it was mysterious.

Later, throughout the centuries, Hippocrates's followers gave the body's curative powers melodious Latin names: *vis vitalis, vis mediatrix, vis essentialis.* But by the late nineteenth century, the quest to solve the mystery of immunity, along with the rest of medicine, had entirely shifted into the materialist sphere. Physicians had completely abandoned the concept of the inner healing powers that had once dominated medicine. In fact, the very notion of such powers had become anathema. After all, no one had ever managed to observe them, let alone measure their strength. Only the homeopaths in Europe and America, scorned by the medical establishment, still spoke of a "vital force" that warded off disease. Any mainstream physician worth his stethoscope had turned for answers to the palpable, material world.

It's possible that Kleinenberg saw the irony. Metchnikoff, a fanatic materialist, was now reviving the "inner powers" idea after having observed the first material evidence of a curative force. It might have even occurred to Kleinenberg how close Metchnikoff was to Hippocrates geographically at the time of his discovery—a breeze of a sailing trip separated the two Mediterranean islands of Sicily and Kos, the birthplace of Hippocrates. Still, more than two thousand years had passed since the time of Hippocrates before the enigma of immunity began to lift.

Encouraged by Kleinenberg, Metchnikoff went on to inject every possible material—goat milk, drops of human blood, carmine or indigo dyes, chunks of cooked peas, sea urchin eggs—into a variety of transparent animals—starfish larvae, marine worms, sea squirts. Invariably, the animals' mobile cells gobbled up the invading particles or at least surrounded them, holding them up.

After about three months an incredibly distinguished visitor, who was unusually qualified to appreciate these findings, was sent Metchnikoff's way as if by heaven—or rather, earth. The squat Mount Etna, a restless volcano south of Messina, showed signs of awakening. As always, whenever a volcano promised to erupt, people within hundreds of miles hopped on carriages or trains to get close to the groaning mountain. Incandescent lava spewing from the bowels of the earth offered the best show on the planet. Among the eager spectators was none other than the über-famous Berlin physician, statesman, and public health reformer Rudolf Virchow, who had revolutionized pathology by pulling existing theories about cells into a unified system. In March 1883 he was in Italy nursing his health in the warm climate when he got the news about Mount Etna, which Johann Wolfgang von Goethe, whom he admired, had scaled about a century earlier. Changing his travel plans, Virchow hurried to Messina.

More than two decades had passed since Virchow's cell theory of disease had sparked Metchnikoff's childhood dream, the one about making his own great contribution to medicine. Now that this fantasy was coming true, Metchnikoff got the unique opportunity to discuss it with the person who had once inspired it.

When they met in the home of a Messina university professor, Virchow agreed to stop by Ringo to look at Metchnikoff's experiments. It is easy to imagine how anxious Metchnikoff was in expectation of the visit, especially if he was aware of Virchow's reputation for being critical and cold.

A small, lithe man who, in contrast to Metchnikoff, never seemed in a hurry despite a staggering record of achievements in different fields, Virchow, then in his sixties, had an authoritarian air that had earned him the nicknames "pope" and "pascha." Even his approval was said to take the form of "icy enthusiasm." But in the case of Metchnikoff's findings, Virchow had a personal reason to abandon his usual iciness. Critics of his cellular pathology had started arguing that it had outlived its usefulness. Metchnikoff's discovery furnished him with a fresh banner for drawing attention to cells, this time as carriers of immunity. So even though Etna never came through (there is no record of it erupting that year), the encounter with Metchnikoff more than made up for Virchow's disappointment.

At the end of April, having returned to Germany, he made the very first public presentation of Metchnikoff's immunity hypothesis, reporting excitedly to the Berlin Medical Society, "Recently, thanks to a lucky chance, I had the opportunity to learn about the truly amazing experimental observations of one of our most famous zoologists, Professor Metchnikoff, whom I met in Messina." Reading these lines in the medical weekly *Berliner klinische Wochenschrift*, Metchnikoff was elated to receive a blessing from the patriarch of cells. Throughout his life, he remained grateful to Kleinenberg and Virchow for their early support.

But Virchow's advice was to proceed with caution. Metchnikoff was encroaching dangerously on the pathologists' holy of holies—inflammation—a centerpiece of pathology texts. Most alarmingly, he was making claims that ran contrary to commonly accepted views: he argued that wandering cells deliberately moved to the sites of inflammation to protect and heal.

"Pathologists believe and teach just the opposite," Virchow told him.

Indeed, pathologists thought the cells were the causes of inflammation. The consensus was that cells leaked from blood vessels and then disintegrated, leading to the swelling, redness, and other hallmark signs of inflammation. When the white blood cells contained microbes, which they frequently did, they were thought to spread the infection. In other words, the common view was that inflammation was passive and harmful. Metchnikoff, on the contrary, was saying that inflammation, in its essence, was active and curative.

When the weather in Sicily started getting too hot, Metchnikoff moved with his family to the town of Riva on Lake Garda, beneath steep cliffs on the southern edge of the Italian Alps. While Olga and her siblings explored the mountainous environs, he prepared a paper about the newly discovered functions of wandering cells. Cautiously, as Virchow had advised him, he stated in its conclusion that cellular digestion "also plays a protective role against harmful bodies formed inside the organism or penetrating it from the outside." He need not have been so cautious. We know today this digestion is an ancient immunity mechanism that helped ensure the survival of species throughout evolution.

On the other side of the Alps, in Vienna, his next stop on the way to Russia, Metchnikoff enlisted the help of colleagues in coining respectable terms for his cells using ancient Greek, which was not part of his polyglot repertoire. The voracious cells became *phagocytes,* from *phagein,* "to eat," and *cytes,* "cells." The process of devouring itself was labeled *phagocytosis.*

Now the new hypothesis was ready for publication. Metchnikoff submitted his paper about phagocytes to a Viennese zoological journal and headed back to Russia, unaware of the impending storms.

10

CURATIVE DIGESTION

WHEN THE METCHNIKOFFS RETURNED TO Russia, the talk of the empire was the ostentatious Kremlin coronation of Tsar Alexander III in late spring of 1883. The repressive counterreforms the new tsar had enacted after his father's assassination seemed to have quenched the rebellious mood of the preceding two decades. There was no more question of liberalization or restricting the monarchy. During three weeks of festive dinners, balls, outdoor parties, and military parades, Moscow was overrun by thousands of visitors, among them princes and dukes from around the world.

Odessa at the time was preparing for its own landmark gathering, the Seventh Congress of Russian Naturalists and Physicians, which more than six hundred participants from all over Russia and even a few foreign guests would attend. Metchnikoff, one of Russia's most famous biologists, was appointed chairman. "The medical contingent of the congress thought that by this appointment, we would mark the close collaboration between biology and the emerging new medicine," one of the organizers explained later.

On the sunny morning of August 23, 1883, Metchnikoff took the podium in the large assembly hall of Novorossiya University, so familiar to him, with its arched windows and a packed balcony. As always, he was nervous in the first few moments, then became increasingly excited as he spoke, characteristically holding up his left arm to the side to stress major points—and, as he always managed to do, electrifying his audience. This was his historic talk, "On the Curative Forces of the Organism"—the first time he unveiled his theory of immunity in public.

"We often hear that so-and-so has recovered from an illness 'thanks to his strong nature,'" Metchnikoff said. "Indeed, since ancient times, medicine had the notion of a curative force of nature, which protects people from disease." He then cited Hippocrates to this effect and the views of Hippocrates's followers throughout history. But more recently, he continued, "the modern school of general therapy has almost entirely abandoned Hippocrates's *physis.* Only a few textbooks still mention it, but they do so in passing and most reluctantly." Still, how do animals and humans fight off disease-causing organisms? Current knowledge suggests they "must possess a certain capacity to fight [the bacteria], otherwise the entire human race would have long gone extinct."

Metchnikoff revealed his own audacious vision of the curative force. Long before humans appeared, he said, plants, worms, insects, and other creatures were attacked by bacteria. "How do plants and animals resist the onslaught of the powerful, ubiquitous bacteria? I believe this general question can be answered as follows: animals disarm bacteria by eating and digesting them." He then ventured a guess that was to prove an amazingly accurate description of what happens in the human body. "Whether the bacteria penetrate [us] through the lung vesicles, the wall of the digestive tract, or a wound in the skin—everywhere they risk being captured by mobile cells, capable of consuming and destroying them."

Finally, Metchnikoff shared with the audience his stunningly prescient idea that complex animals, including humans, possess "an entire system of the organs of curative digestion." In fact, he came up with the concept of an immune system of sorts. He correctly identified the spleen, the lymph glands, and the bone marrow as playing key roles in immunity. These organs, he said, contain special cells that digest germs. "Whereas such lower animals as sponges have one general mass of digestive cells, higher animals have derived two groups of tissues from the same source: regular digestive organs and a system of medical or therapeutic (or rather, preventive) digestion." People probably differ in their susceptibility to disease, he concluded, because they differ in the strength of "the organs of curative resistance."

These were just the kind of humanity-changing ideas that Metchnikoff had striven for since his youth.

The talk created a stir. "During the congress and for a long while afterward, no matter what the topic was at any medical get-together, the conversation invariably turned to Metchnikoff's new ideas," one Odessa physician recalled later. In the fall of 1883 Metchnikoff was elected to the Imperial Saint Petersburg Academy of Sciences, one of the first in the long line of distinguished memberships he was to accumulate in Russia and abroad. The recommendation letter stated that his report at the Odessa Congress had "brought about a new era in the study of pathological processes."

The letter failed to mention that for the time being, the "new era" was entirely based on a brilliant hypothesis. Metchnikoff had yet to see real-life evidence of his "curative digestion." It was rather astonishing that his hypothesis was rooted mainly in intuition, and even more astonishing that it was true.

He soon found the much-needed evidence in a most unlikely place: a friend's aquarium. It contained transparent daphnia—tiny crustaceans commonly called *water fleas* because of their hopping movements in water—that were infected by parasitic fungi. These creatures may not seem to be the most suitable organisms for solving the riddle of immunity, but they enabled Metchnikoff to witness resistance to disease in a way that is particularly valued in biology: in a living animal, in real time.

When he placed sick daphnia under his microscope, he was rewarded with a breathtaking spectacle: a needlelike fungal spore that entered the water flea's body, puncturing its intestines, was immediately surrounded by mobile cells, the phagocytes. He drew up the results after studying about a hundred daphnia for more than two months. If the spores were too plentiful, they matured into fungi that eventually killed the daphnia; but if the cells managed to digest the spores, which happened in about two-thirds of the cases, the flea survived, saved by its own inner curative force.

This image of the curative force in living creatures was so indelible, it was to sustain Metchnikoff throughout the upcoming "immunity war." Years later, he wrote in an essay on the history of the phagocyte theory, "When this theory was attacked from all sides and I asked myself if I hadn't, after all, gotten on the wrong track, all I had to do was to recall the fungal disease of the daphnia in order to feel I was on firm ground."

Yet even the proud father of the brand-new theory realized that to win over the skeptics among physicians, he had to move on from water fleas to mammals. He decided to infect rabbits and guinea pigs with anthrax, the best-studied germs to date. In this, he was extending an invisible hand to Louis Pasteur, whose anthrax vaccine had revolutionized veterinary medicine some three years earlier. Metchnikoff was determined to show that the immunity granted by the vaccine was due to none other than phagocytosis.

He vaccinated two rabbits and two guinea pigs with weakened anthrax germs in an attempt to make them immune to anthrax. The vaccination worked only in one rabbit; it remained healthy even after being injected with virulent anthrax microbes. The other rabbit and the two guinea pigs died after receiving the virulent injection.

Under the microscope, the blood of the four animals told a beautiful story, just the one Metchnikoff so wanted to hear. In the blood drawn from the healthy rabbit, anthrax germs were safely encapsulated within phagocytes. In contrast, in the blood of the dead animals, the germs floated freely. It was obvious, to Metchnikoff at least, that the healthy rabbit had become immune to anthrax because its phagocytes, trained by exposure to the vaccine, had saved it from the deadly disease.

"We now have more reason than before to propose that the intracellular digestion of phagocytes, which gradually get used to digesting substances they previously couldn't overcome, participates in immunity achieved by vaccinations," he wrote in a scientific paper he sent to Rudolf Virchow in Berlin. His earlier paper on phagocytosis had gone unnoticed because it had appeared in a zoological journal. But the two new reports, on daphnia and on anthrax, were bound to attract attention because in 1884 they appeared in *Virchow's Archive*, the leading medical journal of the time. Aching with anticipation, Metchnikoff would often hurry to the library, flipping impatiently through German medical publications in search of comments on his theory.

He nearly missed the first such response because Olga and her twin sister had been diagnosed with "weak lungs." Doctors advised them to spend the winter in a warm climate. Metchnikoff grew understandably anxious, whisking away Olga, her twin, and their other sisters on a tour of the Mediterranean coast. Only in the spring of 1885, finally catching up with the medical literature while in the Italian city of Trieste, did he see the detailed critique of his theory by the prominent Königsberg pathologist Paul Baumgarten in the weekly *Berliner klinische Wochenschrift.*

It was scathing. When a quarter century later Metchnikoff told the story of his struggles over the phagocyte theory in *Immunity in Infective Diseases,* he stated plainly, "The war was launched by Baumgarten."

Baumgarten took the phagocyte theory seriously—and tore it apart just as seriously. He invited the reader to consider, for example, relapsing fever. Most patients recover even though their blood contains freely flowing microbes, unconsumed by phagocytes, "which shows that the living organism can overcome vigorously developing parasites without any help from phagocytes." As for the patients' recovery, he wrote, it happens when the microbes, having reached "the limit of their life's duration," simply die *von selbst,* "in and of themselves."

Particularly hurtful was Baumgarten's claim that Metchnikoff's theory was "teleological," a concept that implies there is a design or purpose behind natural phenomena—in this case, that Metchnikoff supposedly attributed to his phagocytes a humanlike sense of purpose.

This was a grave accusation. All his life, Metchnikoff, along with other Russian nihilists, subscribed to a positivist philosophy of science, which stated that the only valid source of knowledge was observable evidence. Positivism could never be reconciled with such a metaphysical notion as the existence of a design in nature. In effect, Baumgarten was telling Metchnikoff, a high priest of science, that his theory was unscientific.

In the face of the critiques, a lesser man, one who did not believe in the supremacy of science, might have doubted himself or beaten a retreat. Metchnikoff rushed to fight back. The same combativeness that had helped him discover phagocytosis now propelled him to defend it.

Back in Odessa, he threw himself into a fury of new studies. Infecting macaques with relapsing fever, he showed that Baumgarten hadn't found phagocytes in the blood of patients with this disease because in this particular case, phagocytosis takes place not in the blood but in the spleen. "How wonderful that the spleen has been defined as a 'therapeutic organ,'" Metchnikoff's Messina friend Kleinenberg wrote to him excitedly. "It's superb that the significance of a body part familiar for thousands of years has been for the first time understood." Kleinenberg also tried to console Metchnikoff. "You shouldn't be distressed that your work on phagocytosis is not immediately making way for itself," he wrote wisely. "This so often happens in science and can be corrected by one factor only—time."

At this stage, Metchnikoff indeed still believed that convincing his opponents was a matter of a few more experiments. He was determined to prove he was driving medicine forward, not back, in claiming that Hippocratic inner healing was real after all. His major problem was that he lacked an adequate laboratory.

Going back to a university, in Odessa or elsewhere, was out of the question. Metchnikoff tried to work in his Odessa apartment, making the best of the sponges and starfish of all colors that filled his aquarium, but he couldn't conveniently experiment at home on rabbits, guinea pigs, or monkeys. Then, an unexpected solution arrived from Paris.

11

THE PASTEUR BOOM

"BOUND AND HOWLING, OR ASPHYXIATED between two mattresses," in the words of one of Louis Pasteur's collaborators, was a popular image of a rabies victim more than a hundred years ago. The mere mention of rabies, or hydrophobia, as it was once known, elicited irrational fears of savage beasts. Cauterizing the bite wound with a red-hot poker and other attempts at treatment were not much better than the disease itself. That was probably why Pasteur had chosen as the object of his very first human vaccine the horrifically fascinating rabies over less-dramatic common diseases.

In late October 1885, telegraphs throughout the globe were clicking out the sensational news from Paris. A CURE FOR HYDROPHOBIA—DR. PASTEUR ANNOUNCES THE SUCCESS OF THE INOCULATION TREATMENT, proclaimed the *New York Times*, then just over three decades old. In fact, to this day no cure exists for rabies. Pasteur's vaccine was intended to *prevent* rabies—which usually takes a few weeks to erupt—if given shortly after a bite of a rabid animal. Still, the vaccine was nothing short of a medical miracle. It heralded a new era in which all infectious diseases were to be conquered.

In Metchnikoff's Odessa, reports about Pasteur's vaccine struck a chord. "Let's talk about something happier: what's the news about the cholera epidemic in Odessa?" Sholem Aleichem's Tevye the Dairyman says famously, and for a good reason: the human hodgepodge that flowed into this super-busy international port brought with it a disturbingly steady supply of cholera, plague, typhus, and smallpox. So barely three months after Pasteur's announcement, the city council of Odessa was already drawing up plans for a bacteriological station—a facility that, apart from dispensing the antirabies vaccine, would protect the people and cattle of Odessa against a variety of epidemics. Such an establishment, promised the municipal weekly *Vedomosti* in February 1886, would turn Odessa into "an exceedingly important center of knowledge and bathe it with glory, as the first city in Russia to respond to the call of science."

The success of the facility, the newspaper proclaimed, was crucially dependent on the willingness of "our famous scientist I. I. Metchnikoff" to serve as its head. "Our famous scientist," who had not yet met Pasteur, was more than willing. He campaigned for the establishment of the station as if his entire future depended on it. And, in fact, it did—perhaps not his entire future, but at least the future of his phagocyte theory. A bacteriological station could provide him with the conditions for experimenting on warm-blooded animals, something he badly needed in order to defend his theory. Now that public attention was focused on the new vaccine, it was more urgent than ever to show that the immunity it granted depended on nothing else but his valiant phagocytes.

But Pasteur was not quick to divulge his vaccine's secret. The vaccine's first steps had been much more wobbly than reports in the press made them seem. Of the eighty people who had received the shots in Pasteur's laboratory by the end of 1885, three had died. Pasteur argued that without his vaccine, the death toll would have been much higher, but when it comes to rabies, mathematics is notoriously fickle.

First of all, was the dog rabid to begin with? This question cannot be definitively answered unless the animal is captured, which all too often it is not. Moreover, even if the animal is rabid, the outcome depends on the location and severity of the bite; most bitten people never become sick despite receiving no treatment. Finally, the number of suspected rabies cases for any given region is usually small, which makes any statistical analysis of bites by rabid animals unreliable. And the lack of solid statistics on deaths from such bites made it difficult for Pasteur to argue effectively that his vaccine significantly reduced the number of these deaths.

Pasteur's critics argued that his vaccines could kill people instead of curing them. He was accused of being reckless. The word *charlatan*—even *murderer*—occasionally slipped into the critiques. "We've never had as many people die of rabies as we do now that there is a remedy against it," one angry physician thundered at a meeting of the French Academy of Medicine.

Most disturbingly, the skeptics had a point. The vaccine was made of infected nervous tissue, the same "poison" that could cause the disease. Brains and spinal cords of rabbits—which had been infected by means of chunks of brain tissue from rabid dogs—were dried, then mixed with a sterile broth. The longer the drying, the weaker the poison. Injections were given daily starting with the weakest, the one dried for fourteen days, and ending with the tissue dried for only five days or less. In other words, the final shot was close in potency to fresh poison. This meant, the critics argued—not entirely without basis, considering the method was barely a few months old and not yet properly tested—that the person receiving the injection could get rabies from the vaccine. The fact that no one, including Pasteur himself, had a clue as to how the vaccine worked didn't help growing safety concerns.

The rabies controversy threatened to put a shameful end to Pasteur's stellar career. His forty years of service to French science had earned him glory and honors; he had been elected among the "Immortals" of France's Academy of Sciences. Now, if his enemies had their way, he risked being tried for murder. The last thing he needed was for his beleaguered vaccine to be fatally discredited by being misapplied abroad.

Instead of training foreign physicians to prepare the vaccine, he proposed building a large new institute in France, where all the bitten of the world could receive lifesaving injections. Indeed, just before Christmas 1885, four young American boys had arrived from New Jersey. THE NEWARK CHILDREN INOCULATED LAST EVENING. THEY REACH PARIS ALL RIGHT, UNDERGO THE OPERATION BRAVELY, AND THEN GO TO BED AND TO SLEEP, announced the front page of the *New York Times*. The American humor magazine *Puck* ran a back-cover cartoon: "The

Pasteur Boom: High Times for Hydrophobists; Now Is the Time to Get Bitten by a Rabid Dog and Take a Trip to Paris."

Just as Pasteur was unveiling his plans for a new institute before members of France's Academy of Sciences, he received a desperate telegram from Russia: "Nineteen people bitten by a mad wolf. Can we send them to you?" The cable was from the small town of Belyi near Smolensk, a city in central Russia that some sixty years earlier had been burned to the ground by Napoleon en route to Moscow. Pasteur wired back, "Immediately send the bitten over to Paris."

But as always in Russia, no one could budge without approval from the central government. To organize free passports, railway tickets, and a medical escort for the bitten, Bely authorities cabled the district governor, who in turn cabled the interior ministry in St. Petersburg. So it was as late as two weeks after the wolf's rampage that eighteen of the mauled victims, most of them peasants, finally arrived in Pasteur's laboratory. Only one of the bitten—the seventy-year-old priest Father Vasily, wasting no precious time on prayers—had hastened to Paris earlier at his own expense.

With their grimy sheepskin coats and unkempt beards, *les Russes de Smolensk,* as the French newspapers called them, stood out in the motley crowd of the bitten. While they marveled at Paris's palaces and boulevards, a reporter from *Le Matin,* under the headline MUZHIKS IN PARIS, marveled at their innocence: "What surprised them most in Paris was not the bustle of the city nor the beauty of the monuments, but the fact that France doesn't have a tsar!"

Despite the vaccine, three of the wolf's nineteen victims died in the Hôtel-Dieu hospital after coming down with unmistakable symptoms of rabies. To Pasteur's critics, the deaths signaled that the vaccine was dangerous. To Pasteur himself, they meant that particularly in the case of wolf bites, which seemed to be more virulent than those of dogs, the vaccine had to be given as early as possible.

He decided to give a green light to foreign vaccination stations, even sending a generous present to Odessa: three infected rabbits. Nibbling at cabbage leaves in their cages, they were oblivious of their international mission. In their brains and spinal cords, they carried to Russia the rabies virus of French mad dogs.

12

AN ORIENTAL FAIRY TALE

In June 1886, the Odessa Bacteriological Station opened its doors on one of the city's straight central streets lined with harmonious classical facades. It was the first institution not only in Russia but anywhere in the world to administer Pasteur's antirabies vaccine outside of Paris. The newspaper *Odesskii Listok* filed daily reports "to provide accurate information on this entirely new undertaking, which still elicits a skeptical smile on many people's faces."

Inevitably, Metchnikoff was appointed the station's director. Having Olga's income, he characteristically refused remuneration, donating his generous salary of 3,600 rubles a year to the benefit of the station's meager budget. He went out of his way to equip it properly, searching all over Odessa for a heater with a precisely controllable gas supply so that constant temperature could be maintained in the "brain room," where rabbit brains were dried to prepare the vaccine. Since he had no medical training, the vaccinations were performed by two physicians, both his former students.

Before long, dozens of victims of rabid dogs, cats, and wolves from all over Russia, as well as Turkey and Romania, turned up in Odessa—taking advantage of the free tickets granted them by Russia's Southwestern Railways upon Metchnikoff's special request. Exposing their abdomens for the shots, the bitten were almost as frightened of the injections as they were of rabies. "Believe it or not, I'm a military man and no coward, but I was truly scared," one of them told the *Odesskii*

Listok about his first injection. "If I weren't held back by shame, I would have beat it. Like the groom in [Gogol's] 'Marriage,' I was really tempted by the open window."

Despite the general excitement, it didn't take long for the station to run into grave trouble. Pasteur's worst fears seemed to have materialized: a small but significant number of people died of rabies after having been vaccinated in Odessa. Of course, more would have probably died had they not received the injections. But when seven of the first one hundred patients stopped breathing, something had obviously gone terribly wrong.

When the news reached St. Petersburg, the capital's physicians solemnly suggested "avoiding Pasteur's method since it lacked a scientific basis," reported *Novoe Vremia,* Russia's most popular newspaper. The physicians added that the old remedy of cauterizing the wound was safer and at least as helpful, adding that Pasteur's success was due to his "enormous experience" and that his vaccine "could be lethal in inexperienced hands."

Metchnikoff defended the station in every possible forum, stressing that both its physicians had gone as far as to inject themselves with the rabies vaccine in order to demonstrate its safety. But behind the bravado, he was tormented by worry. Each time he got word that a bitten person was dying of rabies in the municipal hospital, he became agitated, dispatching a doctor to examine the patient and then anxiously awaiting the doctor's return. "He tried to deduce from my gait whether the patient had been vaccinated or not," one of his doctors recalled years later in memoirs about Metchnikoff. "If I walked up the stairs with firm, confident steps, he calmed down, inferring that the bitten person had not been treated at the station. But if my gait was rapid and nervous, Ilya Ilyich was out of the door in a minute, already on the staircase showering me with questions."

To everyone's surprise, the station's early failures were soon traced, at least in part, to culinary differences between Paris and Odessa; rabbits in France were larger than in Russia because, considered a delicacy, they had been bred with hares. Being smaller, the brains of the Odessa rabbits used to prepare the vaccine dried out more quickly, losing their potency faster than the larger ones of the French rabbits. Moreover, the vaccines may have been further weakened by drying in the Odessa heat. When the station switched to using stronger vaccines, the results were immediate. There were no deaths among the first one hundred people treated with the stronger vaccine.

But much to Metchnikoff's distress, the new statistics failed to convince the skeptics. A November 1886 discussion of the Society of Odessa Physicians opened with the question "Can Pasteur's emulsion cause rabies in a healthy person?" It never occurred to Metchnikoff that such questioning might, at least in some cases, stem from genuine concern. Interpreting it as a personal attack, he perceived the society's attitude as that of "genuine persecution." At one society meeting, he lashed out against the skeptics with such vehemence that the chairman declared the meeting's protocol unprintable.

Even more traumatic for the beleaguered Metchnikoff was arriving at dead ends in areas unrelated to rabies. Apart from the widely popular training courses in bacteriology for the region's physicians, virtually every aspect of the station's ambitious activity launched under his directorship was met with opposition. Privately practicing physicians complained that the station's diagnostic services for infectious diseases encroached on their territory. When the station initiated the inspection of drinking water and discovered typhus bacteria, municipal physicians refused to accept that this dangerous disease could be transmitted by the water mains.

Worst of all, Metchnikoff's research initiatives were blocked left and right. After ensuring the collaboration of none other than Pasteur himself and being assigned a farm for experiments on cattle plague, Metchnikoff was about to start developing a vaccine against this calamitous disease. Just then, he received a jolt in the shape of a letter from the Odessa Medical Authority suggesting that he was spreading cattle plague in the area. The next blow was a scandal over Metchnikoff's attempt to reduce the population of *sousliki*, a type of ground squirrel that was decimating cereal crops in southern Russia. He proposed infecting these rodents with fowl cholera, but due to fears—unsupported by science—that fowl cholera might turn into human cholera, the Odessa authorities instructed him to halt all such experiments.

In the spring of 1887, at a congress of entomologists in Odessa, an outraged Metchnikoff protested loudly that "various ignoramuses" were "spreading absurdities" about the station. He ended his talk on a desperate note: "Who can conduct fruitful scientific research when there's no certainty that you can work in peace but are under constant fear that due to some denouncement, all your experimental animals might be killed tomorrow, all your work destroyed while in addition you'll be accused of spreading infection."

Most painfully, the main goal for which Metchnikoff had fought when establishing the station—the defense of his phagocyte theory, which he was so sure was to benefit all of humanity—was in serious jeopardy. He managed to conduct only a few, albeit valuable, studies.

Obtaining from a local hospital tissue samples of patients with the acute skin infection erysipelas, or St. Anthony's fire, he was delighted to find plenty of phagocytes filled with the streptococci that cause this infection. In this study, reported in *Virchow's Archive* in 1887, he for the first time divided his phagocytes into two classes, both known today to be essential for the proper functioning of the body.

He called the larger phagocytes *macrophages*, or "big eaters"—a term still in use. Embedded in tissues of various body organs, macrophages are the subject of tens of thousands of modern studies. "The larger and less mobile macrophages play an important role in devouring weakened or dead elements," Metchnikoff wrote, correctly noting that these cells, in addition to fighting infection, perform tissue maintenance. He called the smaller, more mobile phagocytes *microphages*, or "small eaters," noting—again correctly—that they were a subtype of leukocytes. Present mainly in blood, they are quick to arrive in large numbers at the site of infection or injury, and it is these cells that form pus after they die.

Macrophages swell to large proportions after swallowing smaller cells or debris. The dot at the end of this sentence, for instance, could fit hundreds of leukocytes but only several dozen "swollen" macrophages. With the help of dyes, Metchnikoff could clearly visualize the inner structure of both types of phagocytes. Optical microscopes at the time were already close in performance to those today, providing high-quality images of cellular structures magnified up to a thousand times.

A bitter controversy ensued in the wake of these studies on the pages of Europe's leading scientific publications. On one side was Metchnikoff, formulating the basic concepts of the emerging science of immunity. On the other were pathologists, defending their discipline against an interloper.

The controversy quickly deteriorated into mudslinging worthy of rival politicians, partly through Metchnikoff's own doing. He was all the more stung by the rebukes because they came from scientists he so greatly admired, but their authority never stopped him from responding in kind. On one occasion, in *Virchow's Archive*, he charged Baumgarten, a stalwart of German pathology, with incompetence. "I had already expressed my opinion about the imperfection of Baumgarten's technique," he wrote. "I believe yet another source of his mistakes is his insufficient familiarity with blood leukocytes."

The retorts of the pathologists seethed with hostility. They could hardly ignore Metchnikoff's remarkably accurate observations, but they took issue with his audacious interpretations. Baumgarten wrote in a lengthy critique in the *Zeitschrift für klinische Medicin*: "Metchnikoff's explanation of leukocyte activity stems from his rich imagination rather than from objective observations." Another pillar of German pathology, Ernst Ziegler from Tübingen, in the fifth edition of his textbook on pathological anatomy—the world's most authoritative resource on the subject—accused Metchnikoff of medical illiteracy: "Metchnikoff's studies have brought no new observations, and his phagocyte theory is a hypothesis that only shows how unfamiliar he is with the basics of pathology." Perhaps the most insulting censure came from a foremost French physician, Jules Rochard. His description of the phagocyte theory: "An oriental fairy tale born in the head of a Cossack."

It may seem puzzling that Metchnikoff's theory provoked such indignation. But the wrath makes more sense if we keep in mind that his theory appeared to revive the ghost of vitalism, the idea that an ineffable "vital principle" animated all living beings—an ancient doctrine that had long been discredited as incompatible with science. Metchnikoff's contemporaries, proud that medicine had become a proper science, felt threatened by anything smacking of vitalism. Such offensive ideas, coming from a country viewed as a scientific backwater, could disgrace their discipline. At that time they were striving to explain immunity in physical and chemical terms, and along came Metchnikoff with his gluttonous cells that evoked loathsome vitalistic concepts. Not only that, he was an outsider. Worse yet, a Russian!

Metchnikoff felt unable to retaliate properly against the critiques that kept coming from Germany. A Prometheus bearing the fire of knowledge to mankind, he found himself shackled to the brick-and-mortar of the Odessa Station. Barely a year into his directorship of the station, he'd had enough. It was time not just to leave Odessa. As he noted unhappily, his efforts to conduct research in Russia were being thwarted by "obstacles from above, from below, and from the side." He was certain that the health of millions hinged on his theory of immunity, which needed an appropriate arena.

In his emotional essay "Why I Settled Abroad," Metchnikoff describes his childhood yearning for science, his early career, his woes at Novorossiya University, and finally, dwells upon his troubled stint as head of the station, complaining that its sponsors "demanded practical results, yet work aimed at achieving such results ran into endless hurdles." He further explained why he turned down a last-minute offer to head a bacteriological institute to be founded in St. Petersburg: "Having learned a good lesson from the Odessa experience and knowing how difficult it is to fight objections that arise for no sensible reason from all directions, I preferred to find a quiet refuge abroad for my scientific work."

13

A FATEFUL DETOUR

An excellent opportunity to look for a research refuge presented itself in September 1887 when, accompanied by Olga, Metchnikoff traveled to Vienna to attend one of the first medical conferences on what today would be called public health: the somewhat awkwardly named Sixth International Congress for Hygiene and Demography. An article about the congress in the *British Medical Journal* declared, "Hygienic piety has been rewarded." Participants discussed strategies that were already saving thousands of lives: disposing of sewage in sanitary ways, keeping drinking water free of germs, and "spreading hygienic principles among all classes of society."

Metchnikoff was to report on the eight hundred or so patients by then treated with the antirabies vaccine in Odessa. But most important to him, the congress was one of the first to invite bacteriologists, so he had an opportunity to meet key players in this burgeoning field. Still very much an outsider among medical researchers, he refrained from reporting on his nascent immunity studies.

Armed with new contacts made at the congress, Metchnikoff headed for Germany, hoping to find a laboratory that would take him in, preferably in "a peaceful little university town, which would be most favorable to scientific work." It was only natural for him to seek a fertile soil for his research in Germany—the scientific superpower he had admired since childhood—but under prevailing political circumstances it was an inauspicious choice, to say the least.

The Franco-Prussian War had ended more than a decade and a half earlier, in 1871, but its echoes were still fueling the intense rivalry between the world's two leading schools in bacteriology at the time—the German, led by the indomitable Robert Koch in Berlin, and the French, led by the flamboyant Louis Pasteur in Paris. The two schools ruthlessly criticized each other's research, leveling mutual accusations that verged on insults. When the entire world applauded Pasteur's antirabies vaccine, the German medical establishment exhibited no need for such a treatment, *vielen dank*. Instead, it drastically reduced the incidence of rabies using its own method: stringent laws for muzzling its dogs.

So for Metchnikoff, still officially head of a station whose raison d'être was producing Pasteur's vaccine, looking for a post at a German university was more or less equivalent to seeking a position within an enemy camp. Moreover, he brought with him his unpopular theory of immunity, already rejected by some of Germany's best medical minds. Predictably, German scientists were in no rush to embrace a foreigner who infuriatingly kept insisting that their distinguished colleagues had been in error. The search for a tranquil spot in a German lab ended in resounding failure.

The one big success of his trip was a detour to Paris to look up Pasteur, the scientific father of the Odessa Station, with whom Metchnikoff had corresponded previously and whom he had been so eager to meet. Apart from admiring the legendary French scientist, Metchnikoff had been heartened by his support. Several months earlier, in an article on rabies that opened the very first issue of the new journal *Annales de l'Institut Pasteur*, Pasteur had referred to the phagocyte theory as "*très originale et si féconde*" (most original and so creative).

Pasteur was delighted to be welcoming a scientist whose work was being vehemently attacked in Germany. He had never forgiven the Germans for their occupation of Paris, not to mention France's subsequent humiliating defeat at the hands of Prussia. After the Franco-Prussian War, he had famously returned the honorary MD degree the University of Bonn bestowed upon him, writing that "the sight of that parchment," which placed his name next to that of the German head of state, was "odious" to him.

Also after the war, Pasteur had launched his merciless "beer of revenge" project—a study of fermentation aimed at ending German superiority on the international brewing scene—and even wrote a book, the 1876 *Études sur la bière,* which he refused to have translated into German. In his mind, Berlin's Robert Koch had sinned against France by volunteering to serve in the Franco-Prussian War as a physician. It was therefore inevitable that when the embattled Metchnikoff finally reached Paris after his failed tour of Germany, Pasteur greeted him with the honors reserved for a soldier returning from the front.

In his midsixties, weakened by a severe stroke, Pasteur was now far too ill to work in the lab. Instead, he busied himself with supervising the construction of his new institute and keeping a close watch on the antirabies vaccinations in the crowded facility hastily set up in the Latin Quarter, not far from his *École normale* laboratory.

Encountering his *cher maître* for the first time in the rabies clinic, Metchnikoff was struck by his feeble appearance. "I saw a frail elderly man of short stature, the left side of his body semiparalyzed, with penetrating gray eyes and gray mustache and beard, wearing a black skullcap over his short hair and a wide cape above his jacket," Metchnikoff recalled some three decades later in his book *Founders of Modern Medicine: Pasteur, Lister, Koch.* "His pale sickly complexion and tired look belied a person who was unlikely to live many more years, perhaps even just a few months."

There was instant chemistry between the two men. They had a great deal in common: both were outsiders storming the world of medicine with radical ideas—the chemist Pasteur with his vaccines, the zoologist Metchnikoff with his theory of immunity—and both shared an excitable temperament.

Pasteur started their meeting by making a statement that worked like balm on Metchnikoff's battle scars. "Even though my young colleagues have been most skeptical about your theory, I have at once taken your side, for I have long been fascinated by the struggle between various microscopic creatures I happened to observe," he said, adding, "I think you are on the right track."

He then flattered Metchnikoff by personally showing him around the facilities where antirabies shots were being given by his staff. Like

Metchnikoff, having no medical diploma, Pasteur was not allowed to administer the shots himself.

The next day Pasteur invited Metchnikoff and Olga to dinner in his sumptuous apartment at *École normale.* Misled by Pasteur's casual manner and unversed in the intricacies of Parisian *bon ton,* Metchnikoff was not expecting an elegant soiree and showed up in a plain black frock coat. "I was in for a great embarrassment when, upon going up the staircase, I ran into ladies in fancy evening gowns and gentlemen in tails," he wrote in his recollections about Pasteur. He was about to rush back to his hotel to change into tails he had worn in Vienna. Just then Pasteur, in a gesture foreshadowing Metchnikoff's harmonious entry into his entourage, disappeared into a back room and reemerged wearing a frock coat.

Seduced by the cordial welcome and the sight of the grand new institute Pasteur was building on the outskirts of Paris, Metchnikoff, on an impulse, asked if he could get an unpaid position in the soon-to-be-completed institution. Pasteur not only agreed at once but also offered to make Metchnikoff head of a laboratory.

But despite this idyllic encounter that resulted in so tempting an offer, Metchnikoff was not yet willing to abandon his German fantasy. Letting go of cherished ideas was not his forte. Marching right back into unfriendly territory, he and Olga boarded a train to visit a few more bacteriological laboratories in Germany. "We went to Strasbourg, Frankfurt, and Breslau, but everywhere conditions proved unsuitable," Olga wrote diplomatically. Still, incredibly enough, just as phagocytes rush to the hub of injury or infection, they then headed straight to the equivalent of enemy headquarters: the laboratory of none other than Robert Koch in Berlin.

14

FAREWELL

At the 1887 Vienna Congress, Metchnikoff had learned from Koch's chief assistant that the Berlin giant was interested in receiving a sampling of tissue slides—glass plates with thin slices of stained tissue fixed upon them—from his latest study on relapsing fever. Metchnikoff volunteered to deliver the slides himself.

A number of Munich bacteriologists who had overheard the exchange tried to talk him out of going. "They were sure I was going to make a fool of myself," Metchnikoff wrote years later. It was widely known that Koch held views diametrically opposed to the phagocyte theory. He believed that bacteria, far from being devoured by phagocytes, were the ones to penetrate the cells, using them as incubators in which they could multiply. The Munich scientists warned Metchnikoff that Koch would deliberately look for evidence contradicting the phagocyte theory, only to declare it wrong on the basis of personal observations. "Naturally, I ignored the warnings and some time later went to Berlin," Metchnikoff wrote.

Naturally. This particular mix of courage, naïveté, and stubbornness was quintessential Metchnikoff. It had given him the nerve to spar with an acclaimed scientist in his very first serious study, at age eighteen. It had enabled him to create a theory that flew in the face of accepted dogma. And it was now turning him into a fearless founder of the new discipline of immunology, not to be deterred by opposition.

Apart from his drive and daring, Metchnikoff simply refused to forgo the opportunity to meet his hero. He had admired Koch for more than a decade, mesmerized first by Koch's anthrax studies that had supplied crucial evidence for the germ theory of disease, then by his discovery of the serpentine microbe that causes tuberculosis. "My feeling of extraordinary respect for Koch turned into genuine veneration when I read his first report about the tuberculosis bacilli," he wrote in his recollections.

At the time of Metchnikoff's visit, Koch was at the pinnacle of an illustrious career. A former country doctor, he had turned into a universally celebrated founder of the field of bacteriology and, in the words of one biographer, a "crusty and opinionated tyrant." In his midforties—only two years older than Metchnikoff—he had recently been appointed head of the new Institute of Hygiene at the University of Berlin, a key position in a key medical field. Scientists and physicians from all over the world hurried to Berlin to receive training from *Exzellenz* Koch, as he was sometimes addressed, having been appointed imperial privy councillor by Kaiser Wilhelm I.

Metchnikoff never forgot the minutest details of his traumatic first encounter with His Excellency. He recalled:

> The man whom I saw sitting at a microscope was aged but not old, with a large bald patch and a thick, broad beard that had not yet turned gray. His handsome face had an air of importance verging on haughtiness. His assistant timidly informed his boss that as agreed, I had arrived to show him my slides. "What slides?" Koch answered angrily. "Haven't I told you to prepare everything for today's lecture? I see that things are missing." The assistant, apologizing meekly, again pointed in my direction. Without shaking my hand, Koch announced that he was pressed for time and couldn't dwell long on my preparations.

Metchnikoff showed Koch slides with fragments of spleen tissue stained with a violet dye, revealing numerous bacteria, most of them inside phagocytes. Glancing at them brusquely, Koch declared that they furnished no proof of any theory of immunity. "Koch's response and his entire manner stung me to the quick," Metchnikoff wrote later.

Having made the effort to come to Berlin, he could not possibly have imagined that Koch would be so cold.

Faced with a growing realization that their meeting was turning into a disaster, Metchnikoff protested that *Herr Professor Doktor* had evidently been unable to perceive the true significance of his slides in just a few minutes. Summoning considerable courage, he then asked for a longer meeting that would give Koch a chance to closely observe the phagocytes harboring bacteria. Such a meeting was indeed arranged for the next day, but to no avail. Forcing himself to be polite, Koch dutifully peered into the microscope only to declare, "You know, I'm no expert on microscopic anatomy. I'm a hygienist, it makes no difference to me whatsoever whether the spirilla lie inside or outside of cells."

It was all predictable, had Metchnikoff been capable of listening to anybody. A Russian zoologist bursting into Koch's hierarchical universe with outlandish ideas was doomed to be brushed off. Besides, as Metchnikoff had yet to learn, power struggles interfered with science not only in Russia. The politically conservative Koch was engaged in bitter turf disputes with the radical Virchow, the patriarch of cells, who years earlier had spurned his anthrax studies. So anything having to do with cells rubbed Koch the wrong way—it reminded him of Virchow.

After his meeting with Koch, even Metchnikoff was prepared to acknowledge that his German ambition, one he had been nurturing for so long, had been nothing but a pipe dream. Packing up his belongings, and his bitter disappointment, he headed back home. It is a testimony to the generosity of his spirit that in years to come, he continued to revere Koch as a scientist, even nominating him for the very first Nobel Prize. He was to host Koch graciously in Paris and, after Koch's death, placed a wreath in his mausoleum on a visit to Berlin. Koch, on the other hand, after their first meeting, had gone as far as to besmirch Metchnikoff in a letter to a colleague. This letter has now been lost, but according to a science historian who had read it earlier, Koch's description of Metchnikoff had amounted to an attempt at character assassination.

With the position in Germany already out of the question, Pasteur was Metchnikoff's only possible ally. Olga summed it up: "The contrast between our impression of Paris and of Germany was so great that Ilya Ilyich no longer hesitated. The choice was made."

Back in Odessa, his German fiasco behind him, Metchnikoff began to relegate the directorship of the Odessa Bacteriological Station to a successor. A few months after the Vienna trip, he took a leave of absence from the station, relaxing at long last with Olga in her family's Popovka estate near Kiev. Now his major concern was whether he would be able to work productively in such a large city as Paris.

He was in the process of making the last arrangements before leaving Russia for good when a different kind of disaster struck. One day in August 1888, he received a telegram urgently summoning him to Odessa. Thousands of sheep had dropped dead as a result of an anthrax vaccination performed by the station.

In the preceding months, the station's physicians had successfully injected nearly fifteen thousand sheep with the anthrax vaccine they had prepared using Pasteur's method. Their luck changed when they were asked to protect the flock of a wealthy landowner called Pankeyev. Emboldened by earlier successes, they failed to test the vaccine on a small number of sheep, immediately proceeding with the mass vaccination. But this time the vaccine worked like virulent poison, almost instantly killing more than thirty-five hundred sheep.

"On the fourth day of the vaccinations, a large expanse of the steppe offered a horrendous sight: only some two hundred sheep, including the sick ones, were standing on their feet here and there; all the others lay dying or decomposing in the scorching sun," reported the newspaper *Moskovskie Vedomosti.* "The wind spread an unbearable stench for several *versts.* So shocking and nerve-racking was the scene that some of those present burst out sobbing. The shepherds tried to beat up the doctors responsible for the calamity."

Pankeyev desperately turned to his neighbors for help with burying thousands of his dead sheep quickly enough to prevent further spread of anthrax, while Metchnikoff and his assistants at the station frantically searched for the cause of the catastrophe. The physicians gave the vaccine in two shots: the first weak, the second one stronger, so as to build up resistance against anthrax. Had they done it the wrong way around, injecting the unprepared sheep with the stronger vaccine first? Had the vaccine inexplicably grown stronger during storage? Had foul play perhaps been involved?

A veterinarian who studied the case suggested that the vaccine might have been contaminated with tetanus, but the issue was never resolved. The incident, known as the Pankeyev affair, has remained a mystery. Metchnikoff failed miserably to organize a proper investigation, though he was still technically the station's director. He no longer felt responsible for its work and was eager to leave practical grievances behind and devote himself to pure science.

One immediate result of the panic was the Ministry of the Interior's decision to suspend anthrax vaccinations in Russia. As for Metchnikoff, the Pankeyev affair served as a sinister bookend marking the closure of the Russian period of his life. It removed whatever doubts he still had about leaving his homeland. In Olga's words, "This painful episode was the last drop that made the cup run over. The decision about our move to France became final."

The Pasteur Institute beckoned in the distance like the promised land. By joining it, however, Metchnikoff was condemning himself to an unusually bloody battle.

III

THE IMMUNITY WAR

15

THE TEMPLE ON RUE DUTOT

IT IS HARD TO IMAGINE a further cry from Metchnikoff's dream of a "peaceful little university town" than late nineteenth-century Paris. When he and Olga moved there in 1888, they found it electrifying. The capital of a booming colonial empire, Paris had just become one of the first cities in the world to adorn itself with electric street lights. With a population of more than two million, it was nowhere near as large as London but a bewildering ten times larger than Odessa, and considerably noisier. The first motor-driven trams were whisking passengers around, terrifying the horses that still pulled most vehicles. And all over the city, Parisians craned necks for a glimpse of the world's tallest-ever structure, the Eiffel Tower, erected for the 1889 Exposition Universelle despite protests from artists and writers against the construction of the "useless and monstrous" iron letter *A*.

The peaceful and relatively prosperous period, in retrospect longingly named the Belle Epoque, was already well under way. Several years earlier, France's republican government had made July 14 the national holiday, restored "La Marseillaise" as the national anthem, and passed a series of laws to uphold such liberties as freedom of the press and the right to convene public meetings. An unusual splurge of creativity reigned in Paris. The year of the Metchnikoffs' move, Degas painted his *Dancers at the Bar* and Renoir his *After the Bath*, Rodin was chipping away at a block of marble destined to become *The Kiss*, Zola was writing *La Bête Humaine*, and Jules Verne was working on *The Purchase of the North*

Pole. The cafés of Paris, freed from the need to obtain the prefect's authorization, multiplied and flourished. Living up to its international reputation as a party town, the city offered entertainment to suit every taste, from the classical plays by Molière and Racine at the Comédie-Française starring Sarah Bernhardt to lighthearted and often bawdy musical performances in a new type of nightlife spot, the café-concert.

Arriving on October 15, Metchnikoff and Olga landed in a small hotel in the Latin Quarter, near Pasteur's old laboratory on rue d'Ulm, within a short walk from the Pantheon. It was a gloriously sunny day, perfect for strolling along the avenues and boulevards, which had been broadened in the preceding decades to ensure the city center could never again be choked off by barricades. The newly constructed, well-lit *trottoirs* had opened the streets to respectable women, decked out in corseted bustle gowns, and attracted flaneurs, or "strollers," then an emblematically Parisian figure of a blasé urbanite. The dandiest of the flaneurs, dubbed "Napoleons of fashion" by contemporary guidebook *Paris illustré,* patrolled the grand boulevards, fixing their lorgnettes on the passersby only to shower with contempt a wannabe "whose clothes, made by an anonymous tailor, have a common cut, whose beard is not irreproachable, whose gloves don't delineate a well-shaped hand, and who doesn't send out the smoke of his cigar with the grace of masters."

Metchnikoff, with his outmoded suit and his purposeful look, was the antithesis of a flaneur. All the effervescence of Paris was totally wasted on him. Whether he walked along the mythological boulevard Saint-Michel or stopped by the nearby Turgenev Russian Public Library, a cultural center for Russian Parisians where he would soon be much in demand as a lecturer, white rats were all he could think about. They had been the subject of an alarming German study, published just before his arrival in Paris, to which he was eager to dispatch a rebuttal.

To his joy—and the chagrin of Pasteur, so reluctant to leave the Latin Quarter—the new Pasteur Institute was erected at the southern end of Paris, away from the hubbub of the city center. At its opening, it consisted of one H-shaped building on rue Dutot, then in a bucolic community built on former truck farms.

There had probably never been a scientific institution created with such popular enthusiasm. Its construction had been made possible by an international fundraising drive that had produced the unthinkable 2.5 million francs, testimony to the world's enormous excitement over Pasteur's antirabies vaccine. The *Times* of London had been proven right in its cynical prediction that "with the whole world to draw upon, there should be no difficulty in obtaining donations, of which many would be justified by genuine gratitude and others by that irregular form of the same emotion which is said to arise from a lively sense of favors to come."

On November 14, 1888, unaccompanied by Olga—ladies were not invited due to shortage of space—Metchnikoff climbed the semicircular stairway leading to the brand-new institute's main entrance to attend the grandiose inauguration. Before the police had cordoned off the street, onlookers had climbed onto neighboring roofs to catch a sight of the distinguished guests, among them none other than Marie-François Sadi Carnot, the engineer-turned-president of the Third Republic, who was greeted upon arrival by a band playing "La Marseillaise" in the courtyard. Graced by the presence of some six hundred ministers, senators, financial moguls, and other illustrious guests, the ceremony took place in the institute's resplendent columned library. It bore the flavor of vindication: Pasteur had prevailed not just over rabies but over all those who had accused him of charlatanism. Metchnikoff viewed the glittering event distantly, still through the eyes of a stranger. "A few eloquent speeches, praising Pasteur, were delivered and, naturally, several *Légion d'honneur* orders were presented," he was to write later, amused by the French fascination with decorations.

Metchnikoff had no way of knowing—nor had any of the others in that room, extolling the new institute with patriotic fervor—that he, a foreigner, would one day temporarily take over Pasteur's mantle of medical celebrity, becoming the institute's most famous researcher. Nor could he possibly know that in a sense, he was never to leave that hall. Later invited to work elsewhere, he replied that he would leave the Pasteur Institute for one place only, the neighboring cemetery of Montparnasse. In the end he didn't even get that far. As he requested, his ashes rest in a red-veined marble urn on a glass-fronted bookcase in that same library.

Matching the proud flutter of tricolor flags adorning the institute's iron fence, the names of its five research departments all referred to various aspects of *microbie,* from the word *microbe,* coined a decade earlier by a French surgeon and adopted by Pasteur. (*Bacteriology* sounded too German to Pasteur's patriotic ears.) Metchnikoff's department was called *Microbie morphologique,* probably reflecting his earlier zoological interest in morphology, the study of form, as a key to understanding evolution. Today an appropriate name for it might be Cellular Immunology, but in studying the role of cells in immunity, Metchnikoff was decades ahead of his time. A scientific journal by that name, for instance, was to first appear in 1970.

Metchnikoff was the first person to move into the new institute, and he found it to be a scientist's paradise. Even the three-story building itself—today home, among others, to the Pasteur Museum—announced that it was meant to be a temple of science, its exterior elegantly ornamented by stone corners and portals, its impressive interior welcoming researchers into palatial, light-filled labs.

As an embattled founder of a new science, Metchnikoff needed more troops, but so much of the money raised had gone into the institute's construction that funds for hiring staff were limited. Luckily, Metchnikoff's need went hand in hand with his penchant for taking people under his wing. Under his command, an increasing number of trainees from various countries started fiddling with microscopes and test tubes. Soon Metchnikoff gathered around him a small army of disciples. Their mission, and his, was to dive deep into the body at the cellular level so as to prove, beyond any doubt, that phagocytes were the curative force to which humans owed their survival.

From the small space originally assigned, Metchnikoff's ever-expanding lab came to occupy the entire upper floor of the institute's southern wing. Today that floor is home to France's national center for the control of papillomavirus and other facilities; a dozen offices and labs line both sides of a corridor. Only a metal plaque above a bulletin board, "*Laboratoire de Metchnikoff,* 1889–1916," testifies to its historic past. But when Metchnikoff worked here, most of this space, apart from his own two rooms, was a single majestic hall with high ceilings, a floor of bright red and white tiles, and tall windows overlooking a verdant horizon. On the sides and in the middle of the hall stood long tables with enamel sinks, microscopes, steam sterilizers, dyes, and flasks for culturing bacteria. It was a far cry from the humble lab at the Odessa Station.

And most important, contrary to everything Metchnikoff had known in Odessa, there were no council meetings to storm, no students to rescue. "In Paris I finally realized my goal of practicing pure science devoid of any political or public activity," he wrote in the essay explaining his move abroad. His restlessness—which in Russia had seemed an inevitable failing of his character—all but disappeared, converted into nervous vitality. At long last, at the Pasteur Institute, he found his scientific home.

Here in this new home, in parallel with his scientific research, he would begin to formulate the optimistic philosophy that in later years became his trademark. This was just as well; he would soon need all the optimism he could harness.

True, working in this bastion of French science had substantial advantages over being a solitary warrior. Not only could Metchnikoff finally benefit from adequate research facilities, he now had behind him one of the world's most prestigious institutions carrying the name of one of the world's greatest scientists. He also made full use of its heavy artillery, the highly respected monthly *Annales de l'Institut Pasteur*. But behind this idyll on rue Dutot was a disturbing truth. His new stronghold was itself part of a much larger confrontation between two hostile colossi, France and Germany, struggling over the dominion of Europe with the belligerence that ultimately ignited two world wars.

Nearly two decades had passed since *la catastrophe*, the war France had lost to Prussia in 1871, but the French still smarted under the defeat, resentment percolating through their lives from the moment they dipped croissants into their morning coffee. It had been France that had declared that war, driven partly by concern that the neighboring German states, Prussia being the largest and most important, were becoming far too powerful. But the war, so to speak, misfired. France suffered a rapid and humiliating defeat, losing Alsace and Lorraine, while the victorious German states were united into an empire by Otto von Bismarck, under King Wilhelm I of Prussia. For the subsequent decades, the French attitude toward Germany was dominated by a lust for revenge.

Germany, for its part, had gained a military victory, but France's insistence on remaining a great power undermined Germany's sense of security. No less irritating was France's undamaged standing as the international trendsetter in art and style. When Germans themselves referred to someone's prosperity or success, they said, jealously, that he or she lived *wie Gott in Frankreich*, "like God in France." In this atmosphere of intense nationalist ambitions, establishing German superiority in science and medicine was essential to showing that Germany deserved to be the true leader of Europe.

So from the moment Metchnikoff became one of "Pasteur's people," he, the perennial outsider, found himself dragged into a nationalist feud. He was doomed to lead his immunity battles to the drumbeats sounding in the background—the echoes of real war. In fact, it was probably no accident that his most vocal critics were to hail from patriotic Prussia. One Hamburg bacteriologist went so far as to openly reveal his bias when writing about the role of phagocytes in immunity: it is something that "we, in Germany, do not quite accept." Still reeling from this remark years later, Metchnikoff protested in his book on immunity, "It's a grave mistake to adopt a nationalist viewpoint in scientific matters."

Olga wrote ruefully, "Like dark clouds racing and clashing in stormy weather, like waves hurrying one upon another—so did attacks and objections against his theory follow one upon another in a rush. An epic struggle began."

16

ENGINE IN THE DARK

White rats were on Metchnikoff's mind for a reason. In September 1888, when he was still packing his *chemodans* in Odessa, the journal *Centralblatt für klinische Medicin* in Germany had published a shocking study performed on these animals. It claimed that immunity to anthrax had nothing to do with phagocytes. The study opened with a direct challenge to a duel: "Most discussed at present in its relation to [immunity] is the phagocyte doctrine of Metchnikoff. I shall permit myself to report on experiments that highlight a purely chemical viewpoint, with an eye to a possible explanation." The author was the brilliant German physician Emil Behring, who later joined Koch's institute in Berlin—improbably outshining his famous mentor and—no less improbably—winning Metchnikoff over as a friend.

What Behring had done was to draw blood from white rats, known for their sturdiness among laboratory animals. He then placed silk threads with dried anthrax spores stuck upon them into the clear liquid—the serum—left over after blood is stripped of cells. In the serum of cows and other animals susceptible to anthrax, such spores tend to grow bountifully into plait-like, intertwined threads. But in the serum of the white rats, which according to Behring were naturally immune to anthrax, the spores didn't grow at all. Behring's conclusion: the immunity was due to the chemical properties of the rats' wheylike, cell-free serum. What a blow to the cellular concept of immunity.

Indeed, Behring's study was the most prominent in the wave of a new opposition to Metchnikoff, a rival theory of immunity developed in Germany. It was a wave of blood. The rivals argued that blood and other body fluids harbored natural protective substances, as yet unknown. The new theory was referred to as *humoral*, from the archaic notion of the body's vital fluids, or "humors."

Suddenly, there were not one but two immunity camps, divided geographically, with no hint of a future peace between them. Metchnikoff, headquartered in Paris, pictured the body as being protected by voracious wandering cells. Scientists in the no-nonsense humoral camp, mainly German and following the lead of Robert Koch in Berlin, portrayed the body as being filled with what amounted to disinfecting liquids.

So Metchnikoff's cherished six-year-old theory faced mortal danger. The initial attacks on it had come from pathologists, who rejected or even ridiculed it but offered no alternative. Now a new, formidable menace came from bacteriologists. "Not satisfied with destroying the [phagocyte theory], these researchers sought to build upon its ruins new theories, capable of offering a better explanation of the phenomenon of immunity," Metchnikoff wrote in *Immunity in Infective Diseases*, adding with characteristic generosity, "I must immediately state that these attacks were much more significant than those of the pathologists, leading to extremely important discoveries."

This frankness was to come much later. At the time, Metchnikoff panicked. He was not lacking in ego. He wanted to be right, and, of course, he did not want years of research to go down the drain. But perhaps most important, the phagocyte theory was his chance to fulfill that childhood dream of improving the lot of humanity. He'd staked everything he had on it. And he wasn't about to go down without a fight.

As always, the new surge of attacks energized him. One of his very first missions in the new Parisian lab was to take up Behring's brazen challenge. Instead of visiting the cabarets of Montmartre or getting to know the Louvre, which didn't interest him much anyway, he spent his first months in the city acquainting himself with white rats. Could it be that their blood serum indeed rendered them immune to anthrax? With the help of Olga, who *would* have much preferred to spend time in the Louvre, he proceeded to inject these rats with deadly anthrax germs.

Metchnikoff's anxiety echoed through the half-empty Pasteur Institute. Resourceful as ever, he organized an international pageant of white rats—from Kiev, Zurich, and eventually Berlin—in his lab. Contrary to Behring's statements, not all the rats were universally immune to anthrax. Most succumbed. Only a few Berliners continued to scurry about the cage. Moreover, he found no straightforward relationship between an animal's immunity and the effects of its blood on anthrax spores in the test tube. In a paper in the institute's *Annales*, Metchnikoff admonished his opponents: "My experiments have shown that it's quite risky to draw conclusions about what happens in the organism from results obtained with fluids removed from the body."

In ignoring the immunity potential of the body fluids, he was committing the gravest possible sin against the nihilist philosophy of his youth. A scientist who respects no authority, he himself turned into an authority—in his own eyes. Why the serum sometimes blocked the growth of the bacteria was a legitimate scientific question. Metchnikoff did not pause to ponder it for too long; he had a war on his hands.

In the summer of 1889, Paris was filled with provincials and foreigners. Throngs exceeding an estimated thirty million visited the Exposition Universelle marking the centenary of the French Revolution. Along with confidence in the French Republic, the exposition saluted progress, symbolized by the Eiffel Tower, whose immense hydraulic elevators lifted visitors high into the Parisian sky. The ever-curious Metchnikoff must have torn himself away from his microscope at least once to take a tram to the fair on Champ de Mars, where he could join the crowd marveling at Thomas Edison's phonographs in the wide-spanned Machinery Hall or put on earphones to listen to a live transmission from the opera in the telephone exhibit. Before returning home, he and Olga could swing by the nearby Café Volpini, in which Gauguin and his artist friends defiantly exhibited their avant-garde paintings in rivalry to the fair's official retrospective of French art.

But for the most part, Metchnikoff belonged in his lab, physically and mentally. The arrows shot from Germany kept him in a state of perpetual nervous excitement. One Munich bacteriologist stated that as a mechanism that "disseminated microbes," phagocytosis was "a mistake of nature." A new edition of an influential bacteriology textbook, *Grundriss der Bakterienkunde*, again evoked vitalism: "The phagocyte theory assumes that leukocytes have wondrous properties, almost real feelings, thoughts and action, a psychic activity of sorts."

As always amplifying each critical comment, Metchnikoff was quick to conclude that his work was not being taken seriously enough, that he was finished as a scientist. And much as he wanted to gain respect in scientific circles, he also wanted people the world over to benefit from the scientific understanding of immunity. Luckily, he was unfamiliar with the notion of surrender. The deeper he plunged into despair, the higher he rebounded with fresh energy. In response to every major German study on immunity, he continued to fire back his own studies, counterstudies, and critiques—more than a dozen in 1889 alone, his first full year in Paris.

In one critique, he hurried to defend one of his most virtuoso studies, which he had performed back in Odessa. Working with frogs, which are naturally immune to anthrax, he had inserted anthrax spores into their bodies inside tiny bags made of the pith of a reed. The wall of the bag let in the body fluids but left the frog's phagocytes out. The result: inside the bag, the spores sprouted into luxurious threads, suggesting that the fluids were no impediment to anthrax. By contrast, when he placed pieces of tissue infected with anthrax directly under the frog's skin, where phagocytes could have a free rein, the germs were indeed swallowed up by these cells. Clearly, Metchnikoff concluded victoriously, in antianthrax immunity "the phagocytic elements play an active and significant role whose value can no longer be denied."

But to his dismay, in Germany *die Frösche* croaked a different story. In Königsberg, students of his dedicated opponent Baumgarten announced that the growth of anthrax spores was much more affected by temperature than by being placed inside or outside the reed bags. Mockingly referring to Metchnikoff's frog study as his *experimentum crucis*, "crucial experiment," Baumgarten argued in the *Beiträge zur pathologischen Anatomie* that the frogs' immunity to anthrax was due mainly to their low body temperature and to "the biochemical disfavor of the medium."

The young science of immunology was being born—on a battlefield—in these acrimonious exchanges on the printed page. The rivalry speeded the research and ensured the highest standards. Each side knew the opposition would closely scrutinize its experiments. Metchnikoff possessed an outstanding gift of observation, describing his experimental results with a precision that could not be contested. But soon a pattern emerged: he and his rivals conducted the same experiments and reached opposing conclusions, each side unsurprisingly finding proof for its own theory of immunity.

A central point of contention between the two camps was whether the phagocytes actually killed the bacteria. Robert Koch's friend Carl Flügge, a prominent bacteriologist, likened the cells to "cleaners" or "undertakers" that picked up dead germs, writing in the *Zeitschrift für Hygiene,* "The phagocytes appear either to be the victims of bacteria that continue their victorious march, or as graves scattered abundantly outside the battlefield after the fighting is over."

Therefore, to show that phagocytes could kill bacteria, Metchnikoff first had to prove that they swallowed live ones. Using an ultrathin glass tube, he extracted phagocytes filled with anthrax bacilli from infected pigeons, placing the cells in a solution that killed them under the lens of his brass Leitz microscope. For more than an hour, he patiently peered into the lens every few minutes until he was finally rewarded with a heartwarming sight: the germs were growing, sticking out of the defunct phagocytes like deadly blue spaghetti. "Using this method, as simple as it is demonstrative, I was able to show in five pigeons that phagocytes capture the bacteria live," Metchnikoff reported triumphantly in the *Annales.* To make sure these bacteria were not just alive but indeed deadly, he injected them into a white mouse, five guinea pigs, and six rabbits. All but one animal soon died. What better proof was there that the pigeon's phagocytes had swallowed virulent anthrax germs? He could not resist injecting some of that virulence into his *Annales* report, accusing German researchers of making "unsatisfactory observations, which might have stemmed from defective research methods."

Nor was Metchnikoff willing to budge from his contempt for the idea of disinfecting body fluids. He was to pay a heavy price for his obsession: he missed the opportunity to climb yet another step up the ladder of scientific greatness. But his obsession was an integral part of his drive. Had he done nothing to excess, would he have become a founder of a new discipline? As for his rivals, weren't they similarly blinded by ambition?

In fact, we can now say that both Metchnikoff and his rivals had discovered different but equally fundamental parts of the immune system. This system faces a formidable task in ensuring our survival. Both the cells and the body fluids participate in this ongoing defense. Had Metchnikoff and his detractors paused to take an unbiased look at the evidence, perhaps they would have realized this. But neither side took that pause.

Those early immunologists—who did not yet call themselves immunologists, as they were only just beginning to invent the discipline—were a bit like car mechanics trying to figure out the workings of a massive engine in the dark. Each side shouted that they'd found the answer to immunity while shining a flashlight on a single part. And neither side, certainly not Metchnikoff, was ready to even consider how all those parts might fit together.

Admittedly, the immunity mechanisms are so complex that it was almost inevitable for the two sides to occasionally obtain conflicting results. German bacteriologists, for instance, claimed that when bacteria causing *rouget,* a disease of swine, were injected into rabbits resistant against this infection, they were destroyed within a mere fifteen minutes—too fast for them to be captured by phagocytes. This proved that the germs were eliminated by the body fluids, they argued. But when Metchnikoff repeated these experiments in his Paris lab, using bacteria of varying virulence, he often found the rabbits to harbor live bacteria many hours after the injection, which would give the phagocytes plenty of time to act.

Alas, the showdown was stacked against Metchnikoff from the outset. His theory of immunity sounded too outlandish. Germ-devouring cells protect us by rushing to the site of infection? No wonder people had trouble believing it. Humoral immunity, by contrast, with its parallels to disinfection, was so much easier to grasp. Much as an iodine solution cleans up an open wound, so the body's fluids supposedly disinfected it from within.

Moreover, Metchnikoff's proffered cells were new to the medical scene. The cellular theory of disease was only some thirty years old. Metchnikoff's position was thus doubly vulnerable. Himself a newcomer to medicine, he had committed himself to upholding an entirely new medical concept. But if he was wrong, a possibility he hardly ever considered, the entire notion of an inner curative force would crumble. Little wonder that he slept worse than ever and that his nerves were constantly on edge.

17

AN AMAZING FRIEND

THE PHYSICIAN ÉMILE ROUX, A gaunt man with close-cropped hair, a goatee, and striking dark eyes, who had been Pasteur's major collaborator on rabies, had initially rejected the phagocyte theory. But it was one thing for Roux to have read about bizarre properties imputed to wandering cells and quite another to meet in person the phagocytes' charismatic champion. The lively, voluble, hand-waving Metchnikoff, who spoke French with a Russian accent that to some resembled "an avalanche of *R*s," was enchanting in his warmth and dazzling in his erudition. He extolled his phagocytes with the conviction of a prophet: they were destined, no less and no more, to rid humanity of suffering by making people immune to disease. Who could resist such a vision? So when Roux changed his mind about phagocytes, this acceptance was doubly precious.

A fellow department head and future Pasteur Institute director, Roux, eight years Metchnikoff's junior, became his closest friend in his new homeland. In many ways, he was Metchnikoff's opposite. He was reticent and ascetic—no one had ever seen him laugh. Metchnikoff, in contrast, was outspoken and jocular. Roux opted for down-to-earth research with medical applications; Metchnikoff was given to flights of fancy. Roux, who had started his career as Pasteur's assistant, harbored the bitterness of an eternal apprentice. Metchnikoff was never intimidated by Pasteur's greatness, happy to develop his own theories next to a living legend.

Exceptionally dexterous and sharp-sighted, Roux was always on hand to help out with Metchnikoff's imaginative experiments. They became collaborators on some of the studies (possibly with ruinous consequences for Roux). Metchnikoff asked Roux to edit his writings in French. He, in turn, translated German journals for Roux. Some four years into the friendship, he wrote to Olga, who was visiting family in Russia, "Roux is practically the only person with whom I talk a great deal and with pleasure."

Metchnikoff and Roux shared a passion not only for science. From the outset, Olga, just turned thirty, invigorated by the move to sizzling Paris, was a crucial link between the two friends. In an early Parisian photograph, she looks striking. Her parted blond hair tightly combed, Olga is straight-backed and curvaceous, sporting a buttoned-up gown with voluminous drapery at the back of her skirt. In the words of his biographer Émile Lagrange, Roux "immediately fell under the spell of the beautiful young woman scientist with elegant Slavic looks, possessing an artistic talent and a literary taste he himself lacked."

As could be expected from an upper-class Russian woman, Olga knew European languages (she spoke French and knew English better than Metchnikoff) and stayed on top of opinion-shaping fiction. Unlike her husband, she tended to be quiet. Even by the yardstick of the nineteenth century, her involvement in his work was extraordinary. Setting aside her real passion, art, she made scientific drawings and assisted Metchnikoff in the laboratory.

Both Metchnikoff and Olga welcomed Roux into their family. "The Metchnikoffs provided this hardened bachelor with the only sensation of a home, both intimate and intellectual, he was to have in his life," wrote Lagrange. The three of them treated each other with utmost care and attention. Since Roux periodically struggled with bouts of bronchitis caused by his chronic tuberculosis, Olga and Metchnikoff worried about his health constantly. Olga told Roux what to eat, made him get sufficient rest, and mended his black jacket when it came apart. Metchnikoff showered Roux with his nurturing, serving him milk every afternoon to help him get over problems with his vocal cords. Roux, for his part, insisted on paying for Metchnikoff in restaurants, admonished him for staying up late and, toward the end of Metchnikoff's life, even addressed him as "My dear Papa Metchnikoff," or simply "My dear Papa," in letters.

Whenever Olga went away, Metchnikoff would update her on the well-being of both men in her life. After spending an enjoyable summer day in the country with Roux, he reported to her that "the weather was magnificent, the air perfumed, and the greenery uncommonly fresh—we were both so sorry you were not with us."

In another letter, Metchnikoff assured Olga he wasn't coaxing Roux into writing to her. "That's yet another fruit of your imagination, that sentence: 'For god's sake, leave Roux alone.' I swear to you by anything in the world that I didn't even once remind him to write to you: both times he himself asked for paper and started writing. And he asked about you every day." Like an overscrupulous chaperone, he soothed Olga by describing to her how Roux went through his mail: "As soon as he saw your handwriting on an envelope, instead of opening the telegrams he first of all started reading your letter, with undisguised excitement on his face."

It is uncertain whether Olga and Roux were physically intimate, and if they were, when that intimacy began. Roux's biographers are vague on this topic. What's certain—and rather amazing—is that Metchnikoff didn't mind their romantic friendship or, perhaps, friendly romance. Not only that, but he appears to have actually encouraged it. Once, after trying unsuccessfully to convince Roux to spend two weeks with them in their summer home, he wrote to Olga, "When you come back, it may be easier to convince him."

If Metchnikoff the husband revealed himself as a content, not to say eager, partner in this romantic arrangement, he at least had the excuse of following a trend familiar to him from the days of his youth. In nihilist circles in the Russia of the 1860s, multipartner unions frequently arose out of the fictitious marriages to which the nihilists resorted for "rescuing" women. In line with the nihilist philosophy, which held church marriage in total contempt, thus-"liberated" *nigilistki* were free to pursue other love interests, occasionally maintaining long-term relationships with both their fictitious and nonfictitious partners.

Several of Metchnikoff's friends and acquaintances had spent a good part of their lives in such three-way marital settings. Even the trendsetting novel of Metchnikoff's youth, the utopian opus *What Is to Be Done?* by Nikolai Chernyshevsky, which heralds the coming of "new people" as carriers of societal progress, is centered around a romantic triangle. The heroine's semifictitious husband valiantly rises above jealousy—viewed as vile possessiveness of another human being—to allow her liaison with his best friend. "When I saw that she not only wanted passionate love but felt passionate love herself, and when I realized that this feeling was directed at a person who was entirely worthy and capable of replacing me, and that this person loved her passionately as well—I was overjoyed," declares Dmitry Sergeyevich Lopukhov, the noble husband.

Metchnikoff himself had never deliberately striven to "rise above" jealousy, but the open marriages he witnessed in his youth must have left their mark, for he didn't seem to mind Olga's closeness to another man. Besides, his openness apparently stemmed from more than mere complacency. In one letter to Olga he hinted that her relationship with Roux relieved him of his feelings of guilt. "I'm so happy when you talk about Roux, as to a certain extent this makes up for my great debt to you," he wrote.

As could be expected from a three-way relationship, there were times when the arrangement seemed precarious. "At a certain time, Ilya Ilyich believed that my happiness called me elsewhere and tried to prove that I had a right to that," Olga wrote in Metchnikoff's biography, undoubtedly referring to the Roux situation. Yet in that same passage, she stressed the resilience of her marriage: "We had our trials, but our friendship and deep affection always emerged from them firmer than ever."

Over the years, Metchnikoff's ties to both Olga and Roux proved their staying power. After receiving the Nobel Prize, Metchnikoff used his privilege to nominate another scientist for the prize only once—to nominate Roux (albeit unsuccessfully). And during the illness that preceded his death, when Roux visited him daily, Metchnikoff kept repeating to Olga with tears in his eyes, "I've always known that Roux was a kind man and a true friend, but only now do I see what an amazing friend he is."

Those words were spoken after nearly thirty years of friendship. Perhaps the secret behind its endurance was that this friendship never threatened the most cherished relationship of Metchnikoff's life: his romance with science. In Metchnikoff's first years in Paris, his wife and his best friend were his most devoted comrades-in-arms in his grandest struggle.

18

VERDICT

NOW IN HIS MIDFORTIES, METCHNIKOFF was going, in a way, through the exact opposite of a midlife crisis. Risking his health and marriage, he hurled himself completely into his greatest battle. "His full energy as a scientist and a fighter was directed to that; this was perhaps the most intense and turbulent period of his life," Olga wrote in her book, revealing poignantly, "Only his intimates knew how much this struggle cost him in vital force."

In their rented apartment across the street from the Pasteur Institute, on 18 rue Dutot, Metchnikoff tossed in bed through repeated sleepless nights. He used his sleeplessness, which plagued him all his life, to devise ever more ingenious experiments. At an hour when most of Paris had not yet woken up, he rushed to the institute, getting there before the line had formed outside the rabies vaccination service. He often stayed in the lab until late evening, with only a half-hour break in the institute's restaurant (which was provocatively called *Microbe d'Or*, or "Golden Microbe"), and collapsed in his bed before ten o'clock. Naturally, Sundays were no exception. "Ever since our arrival in Paris the work in the lab has been in such full swing that I don't have a free moment," he wrote to a colleague in Moscow.

When a mailman in his shiny-buttoned uniform delivered missives from the enemy camp to the Pasteur Institute, these German publications should perhaps have been labeled "Hazardous Substance." Pasteur, for his part, handled them as befitted a French patriot. "When a

German book or brochure was brought to him with the mail, he picked it up with two fingers either to hand it over to me, or throw it aside with great disgust," Metchnikoff wrote in his recollections about Pasteur.

Metchnikoff, on the other hand, seized upon every piece of printed matter from Germany as soon as it crossed the threshold of the *conciergerie.* He mercilessly tore open the journals' covers, anxiously scanning the table of contents, zeroing in on an item of interest and sometimes reading the entire article while standing next to a tall framed window in the entrance hall. Up in his lab he would write notes on the article, forgetting to pass on the journal to the library, to the disgust of the institute's librarian. Everyone at the Pasteur Institute knew that missing issues could usually be found on Metchnikoff's desk or couch.

With the same anxiety, Metchnikoff awaited news from the Tenth International Medical Congress, held in August 1890 in Berlin. Pasteur and his lieutenants had chosen to ignore this important meeting. Metchnikoff also dutifully stayed away, probably relieved not to face his opponents on enemy territory; his ill-fated Berlin meeting with Koch three years earlier was still fresh in his mind. Instead, he and Olga traveled to Russia to spend the summer in the estate of Olga's relatives near Kiev. But rather than using this hiatus to relax in the country, Metchnikoff nervously awaited what he later called "the verdict of an international jury," to be handed down in Berlin in his absence.

Learning that at the very last moment Pasteur had sent a junior institute member, Waldemar Haffkine, to the congress—not to take part in it, but to report back on interesting developments—Metchnikoff implored Haffkine, his former Odessa student, to write to him from Berlin. And on a trip to the hot and dusty Odessa, he went out of his way to get copies of the *Neue Freie Presse,* scouring it for reports from the congress.

The largest medical gathering ever held anywhere, the Tenth Congress was attended by six thousand or so "medical men," reported the *British Medical Journal.* (They were joined by a total of eighteen trailblazing "lady physicians." Congress organizers had announced, with enlightened smugness, that "they were not morally at liberty" to deny membership to these women.) As was the case with other international medical congresses in the nineteenth century, the meeting served as a prime opportunity to show off the host country, much as happens today with the Olympic Games or the World Cup. Berlin had been abundantly embellished for the occasion. The street leading to a huge circular building called the Circus Renz, containing an enormous lecture hall, was richly decked with flags, wreaths, and festoons of laurel. "As one entered the hall, the scene was dazzling," the *Journal* wrote. "Daylight was quite shut out, and the vast expanse of the amphitheatre was flooded with electric light. Row upon row of ladies and gentlemen, many in evening dress, a large number in uniform, a few in academic costume, rose close-packed to the roof. . . . The picked men of Europe and of all the continents are here."

Invisible amid the glittering crowd, the undecorated Haffkine concentrated on every word pronounced at the tribune. A few years later, he was to become one of Metchnikoff's most famous protégés. Developing vaccines against cholera and the plague, he applied them in India, where these diseases were rampant, earning hero status in that country. Today the prestigious Haffkine Research Institute in Mumbai still wages the never-ending war against germs. But at the Berlin meeting, Haffkine was resigned to the role of a Pasteur Institute spy, filing battlefront reports on the lecturing luminaries.

A colossal statue of Aesculapius seated on a golden throne provided the speakers with a majestic backdrop. Even the August heat didn't interfere with the fanfare of the opening ceremony, in which Rudolf Virchow, his chest loaded with stars and crosses, assured the audience that contrary to popular belief, the Germans were "genuine supporters of peace." Immediately afterward came the part that Metchnikoff, tensely holidaying in Russia, was awaiting with bated breath.

Not enough was yet known about immunity to warrant a special session. Still, recent attempts to uncover the body's inner defenses had so enthralled the medical community that the first two lecturers, medical giants who addressed the entire congress, both brought it up.

Greeted by prolonged applause, the first to take the stand beneath Aesculapius was the famed British surgeon Sir Joseph Lister, a tall, patrician-looking man with Victorian sideburns, later to be known as the 1st Baron Lister of Lyme, or simply Lord Lister. Some two decades earlier, he had rescued surgery from the Dark Ages by insisting that surgical tools be disinfected and wounds be kept sterile. He had been inspired by Pasteur's research on fermentation, years before Pasteur himself established a definitive link between germs and disease.

With that same farsightedness, Lister had from the outset enthusiastically embraced Metchnikoff's phagocyte theory. It didn't hurt, of course, that Metchnikoff had joined the Pasteur Institute: Lister was a Francophile and an admiring friend of Pasteur's. Prior to the congress, he had hesitated over giving Metchnikoff's theory the full weight of his support, even going to think it over amid the white cliffs of England's southern shore. He later wrote to his brother, "For the first few days I could not rest satisfied without reading some articles regarding Metchnikoff's phagocyte theory, sufficient to make me feel sure that I was not going upon a false basis in making that theory a chief point in my address." Agonizing until the last moment, Lister ended up writing his lecture on the train to Berlin.

None of those hesitations came through in his talk. Under an incongruous title, "On the Present Position of Antiseptic Surgery," Lister delivered the very first presentation of the phagocyte theory at an international conference. He argued that this theory helped explain the body's fight against infection—for example, in certain forms of surgery. "It has long been very evident that the living tissues exerted a potent influence in checking bacteric development in [certain] wounds; but what was the nature of that influence? This used to be an enigma, but now receives its natural explanation in the phagocytic action of the cells that crowd the lymph soon after its effusion," he said.

The second lecturer that afternoon was Metchnikoff's implacable enemy Robert Koch. He spoke in a thin, reedy voice, his sentences interrupted by endless *ums* and *ers*, but his words carried unparalleled authority. According to a biographer, a Koch myth reigned in medicine: "The great Koch could do no wrong; if the great Koch said it, it was true." With a few remarks, he could make or break a new theory.

Halfway through his overview of bacteriological research, Koch delivered the blow that made Metchnikoff's worst fears come true. "Much ardor has been shown in working out the nature of immunity," he said. "It is true that this question has not yet been brought to a true conclusion, but the view that we are dealing with a kind of struggle between invading parasites on the one hand and devouring phagocytes on the other is steadily losing ground." Finally, he pronounced his harsh judgment: "It is most probable that chemical processes in fact play the chief part."

Learning from Haffkine's letters about Koch's lecture, delivered with such authority and in such an impressive setting, Metchnikoff was worried and deeply hurt. Roux tried to soothe him in his letters from France, but knowing that Koch's words were gospel in the world of bacteriologists, Metchnikoff feared his theory would be buried definitively. Besides, the ease with which Koch had crossed out his labors was hard to bear. Although he was to make peace with Koch in later years, the pain stayed with him for the rest of his life.

Koch's address at the Tenth Congress is remembered as one of the greatest blunders in the history of medicine, but not because of his thumbs-down on Metchnikoff's theory. In that same talk, he announced that his lab was about to produce a remedy for tuberculosis. So dreaded was this disease that the news instantly made headlines around the world, much as would doubtlessly happen today if a head of a respected cancer institute were to announce that he was on the verge of a cure for cancer. KOCH'S GREAT TRIUMPH, blared the front page of the *New York Times.*

The usually careful Koch had made the premature announcement under political pressure. For all the excitement about the young science of bacteriology, the only practical things to have come out of it were Pasteur's vaccines. So the German Ministry of Culture and the Kaiser's court had urged him to announce his "cure," called *tuberculin,* at the congress, to show the world who was who in medical research.

Soon after Koch's announcement, Berlin came to resemble a biblical town overrun by the sick yearning to be miraculously healed. Thousands of sufferers, some of them accompanied by their physicians, flocked there from all over Europe, seeking injections of "Koch's lymph." Due to shortages of supply, tuberculin came to be traded on the black market. Unusually, one young Scottish doctor who had dashed to Berlin saw through the hype. Arthur Conan Doyle, who had by then already published two novels about an uncommonly shrewd detective, himself displayed enviable shrewdness in reporting back to the *Daily Telegraph* in London that "the whole thing was experimental and premature."

Tuberculin, for a while shrouded in secrecy, was ultimately revealed to be nothing more than an extract of virulent tubercle bacteria in a glycerin solution. It proved useful in diagnosis, but as a therapy it was useless and perhaps even harmful—which it certainly was to Koch's reputation. Yet for at least a few more months, the hysteria continued.

Because of all the commotion, when Joseph Lister himself brought his niece to Berlin to be treated with the new remedy some three months after the congress, in the fall of 1890, it took a week for Koch to see him. While visiting Koch's institute during that week, Lister learned about a sensational new study by Emil Behring, who had joined the institute. Unlike tuberculin, it was to prove a genuine medical milestone. Nobody could have predicted at the time that it was going to be Metchnikoff's undoing.

19

THE SOUL OF INFLAMMATION

Stopping over in Paris to see his *cher maître* Pasteur on his way from Berlin back to London, Joseph Lister dropped a bombshell. He told his French friends that Behring and a colleague in Berlin had managed to transfer immunity to tetanus with injections of the blood serum—from an immune rabbit to mice. The newly protected mice blithely scurried about their cages after receiving three hundred times the lethal dose of tetanus poison. Even more impressively, tetanus-infected mice with spastic legs, normally doomed to drop dead within hours, recovered fully after being injected with the same immune serum. Results with yet another disease, diphtheria, were no less spectacular. In the venerable German tradition of providing a Goethe quotation for every occasion, the researchers concluded the paper with a line from *Faust*: "*Blut ist ein ganz besonderer Saft*" (Blood is a very special juice).

The use of the serum meant a step forward from vaccination. Vaccines could prevent but not cure a disease. They were made of weakened or dead germs, which, after being injected into the body, engendered resistance against disease. Behring's serums, on the other hand, not only protected animals against tetanus and diphtheria but also cured them. Besides, this effect could be transferred from one animal to another, suggesting that the serum itself contained a curative power.

A short while later, scientists were to realize that Behring, in fact, had discovered antibodies, at first called *antitoxins*. These biological molecules, now known to be Y-shaped proteins, were the major secret of humoral immunity. They were precisely the mysterious natural substances that scientists in the humoral camp thought to be present in the body's fluids. Exposure to germs, as occurs in vaccination, triggers the production of antibodies, which are exquisitely precise (or, as scientists say, *specific*) in their action. Antibodies against tetanus, for example, are powerless against diphtheria. The serum could confer immunity from an immune animal to an unprotected one because it contained the antibodies that had been produced against a particular germ.

These details were still unknown when Lister excitedly shared the news of Behring's study with colleagues at the Pasteur Institute. Upon returning to London, he declared in a lecture at King's College Hospital, "I suspect that before many weeks have passed the world will be startled by the disclosure of these facts. If they can be applied to man, the beneficence of these researches will be recognized on every land."

In no land could these studies be recognized more than at the Pasteur Institute, where Émile Roux had been intensely seeking a cure for diphtheria. Shortly after Lister's visit, the stunned Pastorians could read about the Berlin study on the front page of the *Deutsche Medizinische Wochenschrift*, Germany's leading medical weekly. "When we received the report of December 4, 1890, the discovery revealed itself to us in all its greatness, even more so than we had thought," Metchnikoff and Roux were to write nearly a quarter century later in that same weekly publication on the occasion of Behring's sixtieth birthday.

As much as they impressed Metchnikoff, Behring's findings also devastated him: they seemed to contradict the phagocyte theory. The transfer of immunity was achieved by the cell-free serum, so by definition this transfer had nothing to do with cells. But characteristically, Metchnikoff didn't pause even for a moment to contemplate defeat. He retaliated in the only way he knew: an all-out offensive. He could not try to cure disease with his phagocytes—phagocyte research would start saving people's lives only decades later. Technologies for growing cells outside the body were more than half a century away. Still, this hurdle did not prevent him from raising the understanding of immunity to a new level.

Behring's discovery affected him like a catalyst that works its wonders on a chemical reaction. Within months of the publication of Behring's study, Metchnikoff developed a new theory to explain one of medicine's most iconic processes, one that is closely connected to immunity. The telltale symptoms of inflammation—redness, heat, swelling, and pain—had already been described in antiquity. Finally demystifying the biological machinery of inflammation was sure to count as a tremendous feat. That was just what Metchnikoff did. In twelve lectures at the Pasteur Institute in April and May 1891, he presented his ambitious "biological theory of inflammation."

Among his most enthusiastic listeners, wearing a dark jacket and his habitual bow tie, was Pasteur himself, always the first to applaud. In those lectures, Metchnikoff built up what he called "a genealogical tree of inflammation," from single-celled creatures to humans. Throughout the tree, he argued, phagocytes digest living or nonliving irritants entering the body. In more complex animals that have blood vessels, the phagocytes, along with other leukocytes, have a fast conduit for reaching the target. "By means of the blood-current the organism can at any given moment send along to the threatened spot a considerable number of leukocytes to avert the evil," Metchnikoff said. The cells didn't simply leak from damaged blood vessels, as pathologists had claimed. On the contrary, "the leukocytes themselves proceed toward the injured spot instead of passively filtering through a vessel wall," he argued.

Metchnikoff then proudly proclaimed that his theory was "based on the law of evolution. Those of the lower animals that were possessed of mobile cells to englobe and destroy the enemy, survived, whereas others whose phagocytes did not exercise their function were necessarily destined to perish." In the preface to the English edition, he declared, "I have indeed dared to put forward a new theory of inflammation, only because I felt that I had Darwin's great conception as a solid foundation to build upon."

In those lectures, Metchnikoff captured the very soul of inflammation.

His book *Lectures on the Comparative Pathology of Inflammation*, published shortly after the talks, is still widely regarded as a classic. As science historian Arthur Silverstein stated in the preface to a reprint edition, Metchnikoff's work "helped to set the stage for our modern understanding of the nature and significance of the inflammatory response." In modern terms, his view of inflammation was overly simplified, but he was right on the mark in his portrayal of inflammation as a process of active healing. This knowledge was to be crucial to medicine for understanding and treating numerous inflammatory diseases.

But if inflammation is so good, why is it often so bad? Why does it sometimes cause grave damage to the organism? It's as if the body's police, rather than keeping the peace, were themselves occasionally rioting. Today inflammation is thought to contribute to virtually every disease process in the body, abetted by cells that secrete inflammatory substances, among them the phagocytes now dubbed "Metchnikoff's policemen." How can a healing process lead to so much harm?

It seems that for the survival of the organism in the course of evolution, it was more crucial for inflammation to be fast than to be precise. Possibly as a result of that, its control mechanisms, such as those that tell it when to stop, are not as effective as those that set it in motion. Pain, for instance, one of its classic symptoms, is essential for healing because it signals the body to provide rest for the injured organ. But if its controls go awry, it can overstay its usefulness, as it often does.

This idea of "the extension of the normal to the pathological," according to modern philosopher of science Alfred Tauber, was a novel notion that had taken root in nineteenth-century pathology. Metchnikoff's theory of inflammation arose from this idea: disease differed from health only in terms of intensity. "In this setting, Metchnikoff offered a novel concept of health," Tauber wrote in his paper "The Birth of Immunology." Traditionally, health had been perceived as a stable state, one of balance. Metchnikoff's approach was more dynamic. Inflammation was a way of restoring health. If the restoration went too far, it could result in disease. For Metchnikoff, Tauber wrote, "health was not given, but was actively achieved."

In his lectures on inflammation, Metchnikoff pointed the way to actively achieving health. "The curative force of nature, the most important element of which is the inflammatory reaction, is not yet perfectly adapted to its object," he said in the final lecture. "The phagocytic mechanism has not yet reached its highest stage of development. . . . It is this imperfection in the curative forces of nature that has necessitated the active intervention of man."

Shortly after his series of lectures, well-deserved laurels arrived. In June 1891 Metchnikoff was awarded membership in the Academy of Natural Sciences of Philadelphia in the United States and in the Cambridge Philosophical Society in Great Britain, as well as a doctoral *honoris causa* degree from the University of Cambridge. The Latin oration delivered upon the conferment of the degree extolled him for "*generis humani saluti novum in dies affertur incrementum*" (introducing many new advances for the health of humankind).

But in Germany, his theory of inflammation still won him only "a renewal of the objections to the theory," as Metchnikoff himself lamented in the preface to the English edition of his *Lectures*. The acclaimed pathologist from Tübingen, Ernst Ziegler, stated in a new edition of a medical encyclopedia that Metchnikoff's view of inflammation as a combat between phagocytes and bacteria was "not even permissible as a poetical conception."

20

UNDER THE SWORD OF DAMOCLES

"I'M PRETTY NERVOUS, BUT RELATIVELY speaking not so much," Metchnikoff admitted in a letter to Olga upon arriving in London together with Roux in August 1891. They were to take part in the Seventh International Congress of Hygiene and Demography, which was narrower in scope than the Berlin meeting the preceding year but a momentous event in its own right.

Convened under the patronage of Queen Victoria, it brought together more than three thousand delegates "from every important country in the world and many British colonies," according to one British journalist. Her Majesty afforded several foreign delegates, among them Roux, a personal reception in the royal summer retreat on the Isle of Wight. At the opening ceremony in St. James's Hall—then London's principal concert hall—Drs. Metchnikoff and Roux were granted privileged seats, along with the Lord Mayor of London and several dozen leading physicians who, if so inclined, could have a full view of St. James's vividly colored sixty-foot-high ceiling.

There was hardly a more appropriate venue for a meeting on hygiene than overcrowded London. Medicine had already realized that densely populated urban areas—in which poor hygiene was a burning issue—bred disease, turning many maladies into social problems. London, the biggest city in the world, with a population of over 5.5 million, more than Paris, Berlin, Rome, and Vienna combined, was plagued by polluted air. The mortality during its black fogs was as great as in a cholera epidemic. The city had difficulty keeping its water clean and its sewage drainage effective. One Sir John Goode stated in a session on engineering that "the health of London" could not be assured as long as "the sewage of the metropolis was allowed to flow into any part of the Thames without previous purification by the most perfect method as yet known—i.e., by being filtered through land."

According to Olga, "Metchnikoff was delighted that the congress at which he was to give a critical rebuke to his opponents was to take place in England, rather than the hostile Germany." He had already been to England two months earlier, to receive his Cambridge honors, and had come back an enthusiastic fan of the British people. Furthermore, his theory enjoyed unusual popularity in England, perhaps due in part to the all-important backing of Lister.

Such topics as "Prevention of Disease in Growing Towns" and "Enforcing the Ventilation of Public Buildings" were high on the congress's agenda, but the culminating point of its bacteriology sessions was the first-ever public scientific debate on immunity. It was held on August 12 in the Royal Society's splendid mansion on Piccadilly Street and presided upon by Joseph Lister. The debate was a coming-of-age of sorts for immunology, and a contentious one at that. No longer sneaked in among other topics, immunity was finally granted an entire session all of its own.

"Bacteriological workers are at the present moment ranged in two camps, opposed, although in friendly rivalry," the *British Medical Journal* commented euphemistically. As was to be expected, the "friendly rivals" were set apart by a linguistic divide. It was Roux who opened the debate, giving the major talk on cellular immunity. (Metchnikoff probably wasn't the main speaker because, despite his remarkable gift for languages, he was still unsure about his French. His German was better—albeit rusty—but wholly inappropriate for a Pasteur Institute representative.) Speaking for the humoral camp were Munich bacteriologist Hans Buchner and several other German-speaking scientists.

When Metchnikoff took the podium in the discussion, nervous as ever in the first few minutes, he was greeted with cheers. Gradually becoming more excited than nervous and gesturing animatedly, he attacked his German opponents on their own turf, tearing to pieces a study conducted at the Institute of Hygiene in Berlin. With the audience's approval, he ended up talking, in French, for nearly an hour instead of the allotted fifteen minutes.

"Let us take a concrete example and see which of the proposed theories of immunity explains it best," he began. The example, the Berlin study, had shown that the serum of immunized guinea pigs killed *Vibrio metchnikovii,* a microbe named after Metchnikoff by a former Odessa student. But in his own study, conducted in preparation for the congress, Metchnikoff had produced findings of a different sort. Rather than experimenting with the microbe in a laboratory dish as did the German scientists, he had injected the same microbe under the skin of live guinea pigs. He found that the site of the injection turned red, warm, and swollen with leukocytes, which had engulfed the microbes. After elaborating in great detail on all the findings, Metchnikoff concluded victoriously, "In this confrontation of different theories of immunity, a case chosen as the most unfavorable for the phagocyte theory in fact shows that this theory gives the best explanation of the phenomena observed in the resistant animal."

After the immunity session, congress delegates besieged Metchnikoff, wishing to ask him questions. Roux, back in his hotel, dispatched a letter to Olga in Paris: "Metchnikoff is busy showing his preparations and, besides, he would not tell you how great is his triumph. He spoke with such passion that he carried everybody with him. I believe that, this evening, the phagocyte theory is the richer by many friends." Lister concurred. Metchnikoff's talk "carried the large audience almost entirely with it, being enthusiastically applauded," he wrote to his brother.

In the evening after the debate, Metchnikoff was awarded a distinction that some Londoners might have ranked higher than an honorary degree: attending one of the city's most coveted soirees. It was held in the home of Sir Henry Thompson, a urinary surgeon famous, among other things, for having removed the bladder stones of Napoleon III.

For more than thirty years, Thompson gave dinner parties he called "the octaves," to denote a menu of eight dishes and the presence of eight guests, whom he chose no less carefully than the dishes. All male, despite protests from society ladies, they were a who's who of princes, artists, and writers, among them William Thackeray, Charles Dickens, and Arthur Conan Doyle. Metchnikoff attended an "octave" with several British physicians, including Sir Spencer Wells—surgeon to Queen Victoria and, with Thompson, a pioneering advocate of cremation.

Seated at a round table in the dining room hung with classical paintings by Sir Lawrence Alma-Tadema, the diners were served by Thompson's butler—at eight o'clock, of course—in blue-and-white dishes from the host's noted Chinese porcelain collection. A typical menu: oysters, turtle soup, salmon in green Severn sauce, Black Forest venison, pork braised in Bourgogne wine, small fowl timbales with truffles, Argenteuil asparagus, roasted snipes, rum baba, and caviar. The wines, too, were eight in number, among them Chablis, sherry, Hock, champagne served out of a jug or claret, port or Madeira, and brandy after the meal.

Far from flattered by having been invited, Metchnikoff, who resented self-indulgence (he refused, for example, to own a gold watch), was disgusted by the gathering's gourmandism, which might have reminded him of his father's idle hedonism he had so loathed. "Each dish, each sort of wine was commented upon. The wine was first smelled, only then brought to the mouth," he noted disapprovingly in his recollections. "Nothing of the sort ever took place at Lister's," he added further. He praised Lister's lack of pretentiousness, a quality that characterized himself in no lesser degree: "He didn't show off luxury in his home, unlike most of his London colleagues." His cordial relationship with Lister indeed deepened during the congress. "I consider meeting you one of the happiest events in my life," Metchnikoff wrote to Lister a few years later. Lister, for his part, signed a letter to Metchnikoff with the words "your sincere admirer."

The success of the London debate notwithstanding, Metchnikoff was well aware that the battle was far from over. His German opponents showed no sign of laying down arms. Then there were Behring's studies. "In the wake of the London Congress and of several years of research, the phagocyte theory of immunity seemed to have been firmly established," Olga wrote in her book. "Still, the discovery of antitoxins by Behring hung over it like the sword of Damocles." Antitoxins—that is, antibodies—were the putative substances of humoral immunity, then still mysterious and poorly understood.

21

LAW OF LIFE

EVEN IN THE MIDST OF the immunity war, Metchnikoff never forgot that the mission of science was to make the world a better place. And the place he first of all yearned to make better was Russia, even if his strong attachment to his homeland comprised a peculiar mix of devotion and dread.

At the time of his move to Paris, the city had not yet become the most "Russian" of all European capitals—that was to happen years later, in the wake of Russian revolutions—but it already had a sizable Russian contingent: students, political exiles, and Jews fleeing the pogroms. It was the best period ever to be a Russian in France. A month before the Metchnikoffs' arrival, *Le Figaro* had run a front-page article poetically titled "The Russian Soul," in which it had declared, "Russia is à la mode." All things Russian—novels by Tolstoy and Dostoyevsky, gallant counts and countesses, even the accent—were in vogue. The relations between the two countries were warming up toward the signing, in 1893, of the Franco-Russian treaty, which ended the political isolation of a grateful France. The treaty was to elevate the French fascination with Russia to a Russomania. Portraits of the tsar and tsarina were hung in children's rooms. Parisian shops sold billfolds with pictures of the Neva, La Tzarine eau de toilette, and groceries with drawings of scenes *à la russe*, such as polar bears on ice floes.

The welcoming was extended to the tsar's secret police, the imperial precursor of the KGB. It received the support of the French police in spying on Russians in France. According to documents in the State Archive of the Russian Federation in Moscow, Metchnikoff was not under permanent surveillance, but his contacts with exiled rebels occasionally aroused the suspicion of the tsar's Parisian agents. Once in 1894, on a tip from Paris, a secret agent followed Metchnikoff around during his trip to Russia, dispatching detailed reports and encrypted telegrams about his every move to the Department of Police in St. Petersburg. Another secret report from Paris informed Russia's Ministry of Internal Affairs that Metchnikoff had helped create a fund for financing the travel of political exiles back to Russia. He himself did not directly engage in politics, let alone in supporting revolutions, but he could always be counted upon to support Russian mavericks in Paris.

He was particularly quick to come to the rescue of Russian scientists, lending them money and helping them to find lodgings and jobs in Paris. Among the most prominent ones had been his former student Waldemar Haffkine, future developer of the cholera and plague vaccines, who had sent him reports from the Tenth Congress in Berlin. Haffkine had no hope for a scientific career in Russia. Not only was he a Jew who refused to convert, but as a student in Odessa he had helped track down a general who was later assassinated by the terrorist organization Narodnaya Volya. When a desperate Haffkine turned up in Paris, Metchnikoff had managed to arrange for him to be hired as an assistant librarian at the Pasteur Institute until a proper laboratory job opened up.

In fact, a number of Jewish scientists escaping discrimination in Russia found a refuge in Metchnikoff's Parisian lab. It's hard to tell whether this patronage came from his general tendency to defend the underdog or from a feeling for Jews, especially his mother. He always said his mother was "a Jewess" even though her parents had converted. Indeed, in Judaism, Jewish status is maintained regardless of conversion and it is passed on by the mother. Because of that conversion, official restrictions on Jews in Russia or elsewhere didn't apply to Metchnikoff, but his mixed origin might very well have helped to make him an outsider, in life and in science. Still, his own Jewish roots were to him a source of joy. "I ascribe my love for science to my descent from the Jewish race," he told a *New York Times* reporter after winning the Nobel Prize.

Russia's anti-Semitism grieved Metchnikoff, in part because of what it did to the country itself. "Russia has lost many great talents by persecuting the Jews," he once told a reporter. On another occasion, after rejecting an offer to return from France to Russia to head a research institute, Metchnikoff explained in a magazine interview that he wouldn't be able to bring along two of his most dedicated students because "they are Jews, so the doors of the institute would be closed to them."

Over the years, up to a hundred students, researchers, and physicians from Russia were to work at different times under Metchnikoff's guidance in his Pasteur Institute laboratory. Some took part in the bacteriology courses he taught at the institute with Roux. Many returned home to spread the knowledge further. Metchnikoff thus helped to train the entire first generation of Russian bacteriologists—his gift to his scientifically starved homeland, which had not yet created bacteriology departments in its own universities.

In his interactions with others, Metchnikoff displayed a marvelous trait: he treated all people with equal respect, be it a child, a street vendor, or a distinguished colleague, and argued with them with equal passion. He was just as egalitarian in his wrath. Woe to the wretched soul, no matter how high the rank, whom he suspected of sinning against science. Once, he had ordered the belongings of one Prince Peter Argoutinsky-Dolgoroukoff to be emptied into the attic. This aristocratic Russian physician had started research in his Pasteur Institute lab but had been lured away by other attractions in Paris. When the prince eventually did show up again and tried to protest, Metchnikoff threw him out of the institute. "He shouted at me and used coarse language," the prince complained in a letter to Pasteur, hinting at an imminent diplomatic crisis. Metchnikoff had to reassure the worried Pasteur that the incident was not going to undermine the Franco-Russian alliance.

Metchnikoff often despaired at the despotic stifling of freedom in Russia and the undermining of scientific research. But for all his dismay, he never stopped caring about his homeland. Like all people who believe in the object of their love, he looked beyond its current situation, focusing on its potential. Once, pointing to the vastness of

Russia, he told friends, "When this nation becomes well educated, what a contribution it will be to science and to civilization!"

After moving to Paris, Metchnikoff for years continued to refer to France as "abroad." He subscribed to Russian newspapers, ordered books from St. Petersburg, and, despite living in France for nearly three decades, never applied for French citizenship. Even the political scandal known as the Dreyfus affair (the wrongful conviction of Captain Alfred Dreyfus, a high-ranking Jewish officer in the French army, for spying on behalf of Germany) didn't draw him into French politics. The scandal split the entire French society into Dreyfusards and anti-Dreyfusards. Roux was an ardent Dreyfusard, and so were Metchnikoff's other close friends at the Pasteur Institute. Metchnikoff sympathized with the Dreyfusards, as could well be expected considering his acute sense of justice, but he himself never got involved in defending Dreyfus. He was still a Russian in France, as he would be all his life.

Metchnikoff's thoughts were so much with Russia that even in the midst of frantic preparations for the London Congress, he had found the time to write a thirty-page essay, "Law of Life," published in 1891 in the liberal St. Petersburg magazine *Vestnik Evropy*. He had started writing essays on science-related topics in his youth to supplement his income, but with time this writing had developed into a parallel career. It fulfilled his compulsion to broadcast the overriding importance of science. Now he saw an urgent need to counter the views of Leo Tolstoy, even though he greatly admired Tolstoy as a novelist.

Apart from the tsar, no one in Russia was revered more than Tolstoy. The writer was thought to possess such a universal wisdom that a social movement, *tolstovstvo*, or "Tolstoyism"—nurtured by his belief in nonviolence, simple living, and the teachings of Jesus—had formed in Russia in the 1880s. Tolstoy himself not only told people how to live but personally showcased a model lifestyle, making his own boots and taking part in the tilling of his own land.

What worried Metchnikoff was that Tolstoy often turned his wisdom against science. "Scientists can't tell useful knowledge from useless, they study such topics as the sexual organs of the amoeba only because this allows them to live like lords," he once said. "Science deals with anything apart from the questions about what needs to be known, about how one should live."

Metchnikoff feared the consequences of Tolstoy's pervasive influence. "Seekers of truth have not only been avidly reading Tolstoy, but have been implementing his teaching in practice, creating special communities," he wrote in the essay, adding with alarm, "In certain cases, his teaching had caused young researchers to drop science, burn their dissertations, and join communes to start a new life, almost exclusively in the sphere of physical labor."

Not only that, Metchnikoff was disturbed by Tolstoy's preaching for a particular kind of harmony with nature. Tolstoy repeatedly made the incongruous claim that for each living being, animal or human, "fulfilling the law of life" meant following his or her own "natural" destiny predetermined by anatomy. Just as a bird had to fly in order to be happy, he said, so a man had to engage in physical work, whereas "an ideal woman" had to give birth to as many children as possible. (His own wife, having given birth to thirteen, must have come close to his ideal.)

Metchnikoff refuted Tolstoy's zoological claims. "You don't need to be a zoologist to know that some birds don't fly at all, such as ostriches, cassowaries, and penguins," he wrote, further mentioning a species of South American ducks that use their perfectly developed wings only to hover above water. Nature was constantly evolving, he argued—something the fervently anti-Darwinist Tolstoy refused to believe—so the very notion of "natural" was in flux too. He scoffed further, with his habitual bluntness that often offended people. "Excessive exercising of muscles evidently didn't leave [Tolstoy] the time to become acquainted with many scientific questions on which he often passes harsh and entirely erroneous judgments," he wrote.

Besides, for years Metchnikoff had been arguing that harmony with nature was an illusion. In his essays, he took issue with the cult of the harmonious body as perceived by ancient Greeks. He also objected to calls for a return to nature, such as those by Enlightenment philosopher Jean-Jacques Rousseau, who had opened one of his treatises with the motto, "Everything is good as it leaves the hands of the Author of things; everything degenerates in the hands of man."

Metchnikoff, in contrast, in his early essays had been pointing to what he saw as innumerable "disharmonies" inherent to human nature. Many of them sadly hinted at sources of his own suffering: the extreme sensitivity of children due to the long time it takes human babies to mature compared to young animals; the emergence of sexual desire long before marriage becomes appropriate; and the need to manage in society even though humans had evolved from animals that were solitary or living in small groups. In the Tolstoy essay, he pointed to yet another "disharmony" that was later to form the core of his theory of aging: the multitude of "useless" organs in the human body, among them the appendix.

Metchnikoff believed that instead of worshipping nature, humans had to take progress into their own hands. "All organisms apart from man move toward perfection unconsciously. But we must perfect ourselves consciously, with the help of science," he had once jotted down in a notebook in his youth. Olga said that he called himself a "progressive evolutionist"—"He believed that only conscious inner development could lead to genuine progress." Later in life, he was to shape this worldview into an entire philosophy that optimistically affirmed the human ability to alter nature.

In the Tolstoy essay, he defined the rationale for setting visionary goals. He wrote, "A sharp divide characterizes all vital issues: on the one hand, nature with all its negative aspects inherited from animal ancestors—and on the other hand, a powerful, relentless striving for an ideal, for a better future. . . . Tireless scientific labor, coupled with this striving, is bound to bear ample fruit."

22

BUILDING A BETTER CASTLE

It was to save children that the German Emil Behring and the Frenchman Émile Roux were trying to battle diphtheria.

Diphtheria was then one of the most dreaded killers, with a pernicious predilection for the very young. Starting surreptitiously with a sore throat, it choked its victims as if by an invisible hand. Their throats filled with a leathery membrane that earned diphtheria its nickname, "the strangling angel of children." To prevent their young patients from choking, the helpless doctors sometimes inserted a silver breathing tube through a cut in the neck into the child's windpipe. Still, more than half of the children never returned home from the hospital. Nurses didn't last long in diphtheria wards, finding them too depressing.

In working on diphtheria, as in the rest of his research, Roux took no part in the immunity war. If he ever got involved in the immunity debate, as he did at the London Congress, it was to support Metchnikoff. Behring, on the other hand, was a major player in the humoral camp. He became Metchnikoff's most unlikely friend.

They probably met for the first time at the London Congress. Emil Adolf Behring cut a towering figure, soon to be matched by a towering status in the medical world. A former army doctor with the demeanor of a Prussian surgeon-major, he was a brooding, domineering man with a volcanic temper. Institutionalized on several occasions with debilitating depression, he is thought to have suffered from a bipolar disorder. He got up at four every morning, launching his day with a breakfast of fried beef or a large schnitzel, and spent years searching for what he called "inner disinfection," protecting the human body against infection using chemicals—to quote his own analogy, much as "a ham was conserved against putrefactive bacteria by smoking." In London, two days after the immunity debate, he delivered a talk, "Disinfection in the Living Body."

Metchnikoff must have been impressed by Behring's dedication, but how could he become friends with a man who in one of his brochures had referred to the phagocyte theory as a "metaphysical speculation," argued that this theory "relied on mysterious powers of the living cell," and announced that it "interfered with treatment efforts"?

For one thing, Behring deliberately distanced himself from the immunity war, criticizing diehards on both sides. "Born theoreticians like Buchner and Metchnikoff with irrepressible optimism seek to incorporate the new facts into their beloved theories," he once said in a Berlin lecture. What mattered most, he proclaimed, was "how we can use these [facts] to the benefit of sick humans." After the London Congress, in October 1891, he had sent Metchnikoff a friendly letter. "I continue to deal with sick people and am further removed from theoretical questions," he wrote. "For me, obtaining therapeutically and immunologically effective bodies is far more important than clarifying how they work."

In other words, Behring was putting his energies into practical matters, effectively withdrawing himself from the immunology battlefield. He was just the kind of enemy Metchnikoff could love. Metchnikoff's response was warm and generous. "I am convinced that you have made a first-rate discovery," he wrote back to Behring, even adding a rare concession. "I believe we can work peacefully side by side. We can mutually support one another, as do phagocytes and antitoxins."

For Behring too, their friendship had special appeal. Paradoxically, he was both a German nationalist and a flagrant Francophile. Pasteur's picture hung in his study next to those of Bismarck and Frederick the Great. So part of Metchnikoff's attractiveness was surely his senior position in the institute headed by Pasteur. And there was more. Through Metchnikoff, Behring sought to get closer to Roux, who was making rapid progress in preparing a potent serum against diphtheria.

As if making a mockery of the Franco-German hostility, the conquest of diphtheria advanced in an almost symmetrical back-and-forth between the two countries. It had been German scientists who in the early 1880s had first identified the slender diphtheria rods, sometimes clustered in groups resembling Chinese characters.

Then came the discovery that marked the peak of Roux's scientific career. In 1888, together with a colleague, he found that diphtheria germs wreaked havoc in the victim's body indirectly: by releasing a toxin. Roux and the colleague tried to generate immunity against the toxin by exposing laboratory animals to gradually increased doses, but this approach failed.

According to one of Roux's biographers, what followed next was a fatal omission: "They envisaged transferring immunity with blood, but Metchnikoff, exclusively focused on phagocytosis, discouraged them, and they abandoned these attempts." The claim does sound like hindsight wisdom, but if true, then Roux paid a dear price for his friendship with Metchnikoff.

In any event, the only thing known for sure is that the next, crucial step toward a diphtheria remedy was ultimately made by the other Emil. It was the seminal research that so impressed Lister and that stunned Metchnikoff and Roux when reported in the 1890 paper. Behring had shown that blood serum could neutralize the diphtheria toxin after being transferred from an immune animal to one that had not been previously protected.

But how did the serum work? It did not contain the weakened microbes of Pasteur's vaccinations, but it obviously somehow disabled the toxin. The mystified bacteriologists raised various possibilities: that the serum harbored "remnants of vitality," "defensive proteids," or protective "alexins," from the Greek for "to defend." Behring himself at first carefully referred to antitoxic activity, not to a particular substance. He once quipped that isolating a pure immune substance made no more sense than trying to concentrate the essence of Rhine wine. But once such natural biological substances, the antibodies, were defined, they were to dominate the thinking about immunity for decades.

The final offensive against diphtheria—turning the serum into an effective therapy—was undertaken in parallel on both sides of the Franco-German divide. To be used in children, the serum had to be produced in large quantities, which could not be obtained from rabbits or guinea pigs. In France, Roux managed to produce highly potent serum in horses. He first injected the animal with weakened diphtheria germs that didn't make it sick but rendered it immune against diphtheria. Then he stripped blood drawn from the immune horse of all cells, leaving only the clear serum containing the antibodies. Each prancing "factory" manufactured three liters of serum a month, guaranteeing sufficient quantities—and for decades inextricably linking immunological facilities with the smell of horse manure.

But it wasn't enough to produce large quantities; the serum had to be reliably strong. In Germany, Behring's remarkable collaborator Paul Ehrlich developed a way of generating standardized serum with a predetermined number of *Immunitätsgraden*, "degrees of immunity." To make sure the animals manufactured antitoxin in sufficient concentration, they were immunized with regularly increased quantities of the diphtheria toxin. The way to mass production of the serum was now open. Ehrlich was then asked to sign off his rights to the royalties from the serum to avoid a conflict of interest, so he could be appointed head of a state-funded institute charged with monitoring the quality of commercial serum. According to his biographers, Ehrlich felt that by maneuvering him into this agreement, Behring had cheated him out of his share of the profits. Ehrlich stopped collaborating with Behring but maintained an interest in immunity and was soon to become Metchnikoff's major menace.

Finally, there was enough serum to start filling hospital syringes with the straw-colored lifesaving fluid. It sometimes caused a rash or other unwanted reactions—after all, it hadn't come from a human being—and it didn't always work. But when it did, it brought joyful tears to the eyes of incredulous doctors and nurses. Within hours of the injection, blue-faced, gasping babies came back to life, their fevers subsiding, their throats cleared of the suffocating slime.

Very soon, doctors from St. Petersburg to Seattle were injecting horse serum into their diphtheria patients. Within a few years the therapy slashed mortality from diphtheria almost in half, from nearly two-thirds to less than one-third of the cases, and by 1910, to about one-tenth. In the first two decades of its use, it is estimated to have saved more than a quarter-million lives around the world.

Serum therapy was ultimately to prove ineffective against most other infections. Even with tetanus, against which it had produced spectacular results in mice, it was found to work best only as a preventive vaccine. But at the outset, it generated a tremendous sense of triumph. For the first time in history, physicians could cure an infectious disease. And this happened about half a century before the advent of antibiotics, which today are commonly—and mistakenly—viewed as the first successful treatment for infectious diseases.

Remarkably different fates awaited Behring and Roux in the wake of their serum success. The therapy transformed Behring from an obscure doctor toiling in Koch's shadow into one of the world's most highly respected and wealthiest physicians. He was deluged with honors, including the very first Nobel Prize in Physiology or Medicine in 1901.

In France, the serum, just the kind of miracle the French people were expecting from the House of Pasteur, brought an outpouring of donations, saving the institute from financial trouble. A Pasteur Institute hospital was erected across the street—today rue du Docteur Roux—and the government provided a subsidy for the maintenance of more than a hundred horses so as to supply all of France with the serum.

But there was no chance at all that Roux would derive personal financial profit from the serum. For one, a French law on patents prohibited profiting from medicine. The antirabies shots, for example, were administered for free. But even if a loophole could be found in the law, getting rich off medical research was considered *mauvais ton*, "bad form," in France. Pasteur, for one, was contemptuous of German medical entrepreneurs. Besides, Roux was so personally ascetic that had he gotten a share of the profits, he would have surely donated it. (Once, a student of Metchnikoff's was surprised to discover that Roux had paid his monthly dues for lab materials at the Pasteur Institute. "You're young, you need the money, and I'm old, I don't," Roux explained. He was thirty-five at the time.) As for the Nobel Prize, Roux was nominated but became a "highly qualified loser," in the words of a Swedish science historian. The Nobel Committee felt Roux's isolation of the diphtheria toxin was less important than Behring's subsequent discovery of the immunity transfer, while his hospital work was not prizeworthy at all.

The serum did turn Roux into a reluctant celebrity in France. Patriotic French press hailed him as the discoverer of serum therapy. *Le Figaro* bragged in a front-page article titled THE CURE OF CROUP AT THE PASTEUR INSTITUTE, that because it had originated with Roux's discovery of the toxin, this therapy was "an exclusively French undertaking." The scrupulous Roux protested repeatedly against getting all the credit, insisting that Behring share with him all his important French prizes and honors. Still, the remedy itself was known in France as "Roux's serum."

According to one Behring biographer, Metchnikoff, "the linguistically gifted Russian," provided "a bridge" between Behring and Roux, who spoke no German (despite his Francophilia, Behring apparently wasn't fluent in French). This bridge dominated Behring's priorities when he chose godfathers for his sons, using the christenings as opportunities to foster professional and social ties. He had six such opportunities. He gave Roux the honor of serving as godfather to his firstborn, Fritz; Metchnikoff served as godfather to Behring's fifth son, Emil. The godfather of the sixth son, as dictated by tradition, was the Kaiser.

Over the years, Metchnikoff and Behring exchanged dozens of letters and even collaborated in science. Metchnikoff visited Behring in Germany, on one occasion carrying back with him to Paris immunized guinea pigs for a joint study on tuberculosis. Behring, for his part, never missed a chance to visit Metchnikoff and Roux at the Pasteur Institute. When World War I broke out, he reached out to his Parisian friends across the trenches, announcing publicly that he had no intention of breaking contact with Metchnikoff and even sending Metchnikoff a 1915 volume of his collected papers with the admonition, "*Travaillons*" (Let's work).

Even though Metchnikoff initially felt threatened by Behring's transfer of immunity in laboratory animals, he was unfazed by the progress of serum therapy in humans. "The capacity of blood serum to kill bacteria outside the body is a well-established fact, but immunity has nothing to do with it," Metchnikoff announced at one medical meeting. He argued that even if the serum did affect immunity, it was by acting on the cells. "It's much more likely that the antitoxins work by stimulating cellular resistance than by directly destroying the toxins," he wrote on another occasion. Furthermore, Metchnikoff kept arguing that diphtheria and tetanus were exceptions, that the action of the serum against a specific disease didn't explain immunity. He was unaware that this "specificity" was the most remarkable property of the immune response, its ability to counter each threat by exquisitely tailored means. In his dismissal of serum therapy, he failed to notice that this therapy was gradually turning his phagocytes into a museum piece.

On the other hand, Metchnikoff revealed amazing prescience in arguing that if one day the world were to be rid of disease, it would be by strengthening the body's powers of resistance. "Hence the need to learn more about the curative forces of the organism in order to enhance their beneficial action by artificial means," he asserted. It was an ambitious program. He believed that pure science ultimately stood a better chance of ridding humanity of disease than a preoccupation with immediate cures. "It is well known that pure science searches for the truth without worrying about its practical application," he concluded proudly in one review. Serum therapy was no match for this grand vision. At the time that Behring was building a castle-like villa near Marburg with the serum money, Metchnikoff was busy building his own castles—in the air.

23

THE DEMON OF SCIENCE

AS METCHNIKOFF WAS COMING TO terms with Behring's diphtheria studies, he got a chance to apply himself to science that could save myriad lives and benefit mankind. In the early 1890s cholera, a disease well familiar to him from his time at the Odessa Station, posed an urgent threat to the world's population. It was sweeping the globe for the fifth time in the nineteenth century. Leaving its lair in India, it stole onto steamboats and trains. It was carried by pilgrims, merchants, and travelers to Egypt, the southern ports of Europe, and, via Persia, to Russia.

In Russia, a country ravaged by poverty and hunger, the pandemic claimed a monstrous number of lives. In the summer of 1892, nearly three hundred thousand people died of the disease in seventy-seven provinces. A wild rumor claimed that cholera was a cover story. Rather, doctors were poisoning people and burying them alive. Outraged Russians looted or burned hospitals in several Russian cities. At least one "poisoner" doctor was lynched. That same year in Europe, an epidemic carried away more than eight thousand lives in Hamburg. It was the last major European outbreak of cholera—the disease was to be soon bridled in this part of the world by improved sanitation—but no one knew this at the time. Panic spread throughout the continent. People changed and canceled travel plans, and mail delivery was disrupted.

Inevitably Metchnikoff, along with scores of researchers throughout Europe, turned to investigating cholera. He and his immensely devoted assistants resorted to the tried, if ghastly, research scheme: they started drinking glass after glass of water mixed with cholera germs—from the Seine, from the stools of sick people, from a fountain on one of the squares in Versailles. "I've been swallowing *toutes les cochonneries* [all sorts of filth]," Metchnikoff told an amused audience at one medical conference.

In taking such risks, he may very well have been emboldened by the lucky outcome of his self-experiment with relapsing fever more than a decade earlier. Today such practice would be considered criminal when applied to anyone but the experimenter himself, but when Metchnikoff's protégés were drinking their cholera concoctions, laws regulating human experiments were decades away. Not only that, society viewed self-experimenters as heroes. And above all, Metchnikoff personally took the lead. Himself a scientific kamikaze, he could not resist sending his volunteers off on perilous missions.

His goal was to get to the very root of cholera. Was the comma-like germ identified by Koch as the cholera bacterium indeed the cause? Most confusingly, "Koch's commas" were sometimes harmless—they had been found in the stools of healthy people. Metchnikoff's assistants had scooped them out of the Seine at a time when there was no cholera in Paris. And in Munich, Max von Pettenkofer, a highly respected scientist who did not believe that the commas caused cholera, swallowed a whole spoonful—many millions of cultivated cholera germs. So did his students, and none of them became ill.

To clarify the connection between cholera germs and the disease, Metchnikoff asked doctors from all over France to send samples of blood and fecal matter from cholera patients to his lab. "I keep chasing after cholera cases," he reported in a letter to Olga in June 1893. When cholera broke out in Brittany, he tore off to that region. In Paris, he rushed to obtain stool samples from a boat owner who had daily drunk two liters of water from the Seine until being hospitalized with severe diarrhea. Knowing that cheese could sometimes cause poisoning, he traveled to Brie, east of Paris, to test local cheeses for the presence of comma-like microbes.

When one lab worker came down with full-blown cholera after swallowing the germs, Metchnikoff agonized lest the young man might die. But as soon as the worker recovered, Metchnikoff planned more experiments. "The 'psychosis,' as he was to call it later, repeated itself,"

Olga wrote. "Despite the horror he had been through, he decided to experiment on human beings again."

The decision had grim consequences. One of the volunteers, a nineteen-year-old who suffered from epilepsy, developed cholera, then recovered from the disease but died several days later from an unknown cause. Suspecting that cholera might have hastened his death, Metchnikoff vowed never to experiment on people again—a vow he was to keep with cholera, but not with other diseases.

In these risky studies, Metchnikoff, among numerous other findings, managed to confirm Koch's discovery, showing that the germ indeed caused cholera in humans—a tremendously significant achievement in itself. But most important for his own future research, the studies drew his attention to intestinal microbes. These microbes, he hypothesized, could determine whether the person infected with cholera germs would become sick or not. After all, in a laboratory dish, various other microbes could either stimulate or block the growth of the cholera germs. *Perhaps the same thing happened inside the gut,* he thought. "The flora of the human stomach has hardly been investigated, and that of the gut—even less so," he noted in an 1894 paper on cholera. Soon, the *Times* of London reported, "Metchnikoff has invented yet another theory, very curious and rich in appalling possibilities. The cholera germ, he thinks, is only powerful for evil when the native bacilli of the human interior, the flora of the stomach and intestines, as he quaintly calls them, are favorable to its growth."

As the cholera pandemic was winding down, dangers of another kind loomed for the French people. Suddenly, political activism spilled into a wave of violence. Anarchist extremists, seeking to avenge the police brutality during a May Day demonstration, rocked Paris with a series of dynamite explosions. The flare-up culminated with the popular president Sadi Carnot being stabbed to death by an Italian anarchist in June 1894.

But it was a different piece of news that shook Metchnikoff. In the spring and early summer of 1894, his German rivals, having stormed upon cholera with no less urgency than himself, came up with ominous findings. "Just when the theory of phagocytosis seemed to have finally obtained the rights of citizenship, a new discovery appeared to overturn it completely," Metchnikoff was to write in a historical overview years later.

The finding had been made in the headquarters of the humoral camp. Richard Pfeiffer, the imperial-mustached director of scientific staff at Koch's Institute for Infectious Diseases in Berlin, immunized guinea pigs against cholera. He then injected live cholera germs into their bellies and withdrew small amounts of their abdominal fluid every few minutes. Under the microscope, he saw that the bacteria in the fluid had become motionless, lost their comma shape, and within twenty minutes disappeared, like sugar dissolving in water. The finding won Pfeiffer a degree of immortality; this disintegration of the bacteria is still known as the Pfeiffer phenomenon. Since there were very few cells in the abdominal fluid, it was only natural that Pfeiffer announced a victory for the humoral camp. "In the cholera infection of guinea pigs, Metchnikoff's phagocyte theory must definitively be regarded as erroneous," he declared in the *Zeitschrift für Hygiene und Infektionskrankheiten.*

The finding knocked out all of Metchnikoff's defenses in a single blow. He had been saying that diphtheria and tetanus were special cases, unrelated to other diseases. Yet cholera *was* another disease. Worst of all, he had been arguing that the properties of body fluids in a test tube had nothing to do with immunity in the living organism. Yet the Pfeiffer phenomenon was observed in the nearly cell-free fluid of living animals.

By Metchnikoff's own admission, the Pfeiffer phenomenon nearly drove him to suicide. "I was ready to rid myself of life," he wrote.

Luckily, with age, he had learned to turn suicidal impulses to his advantage. Rather than sink into prolonged gloom, he used his bouts of despair as springboards for action. In a series of feverish studies, he injected cholera germs into guinea pigs until finally reaching the conclusion that his eleven-year-old theory was out of danger. He had found a "solution" to the Pfeiffer phenomenon, declaring that the substances that killed the microbes in the body fluids must have been released by phagocytes.

"The pitiful waste of this brainy Metchnikoff's life was that he was always doing experiments to defend an idea, and not to find the hidden truths of nature," Paul de Kruif castigated him in *Microbe Hunters.* The "waste of a life" is a literary hyperbole, but it gets the idea across. Instead of using the opportunity to grasp a full picture of immunity, as resulting from the combined action of body cells and fluids, Metchnikoff saw only his phagocytes. The same relentlessness that had driven him so far in science became his greatest failing when it prevented him from letting go of an idea. As in inflammation, in which harm can follow from healing, this failing followed directly from one of his major strengths.

At the time Metchnikoff was grappling with the Pfeiffer phenomenon, the immunity war was about to reach its peak—at the Eighth International Congress of Hygiene and Demography, held in Budapest in September 1894. Metchnikoff had been appointed honorary president of one of the congress's sections, Etiology of Infectious Diseases, and put up in the fancy Hotel Hungaria ("unbelievably posh, gilded to disgust," he wrote to Olga). In his talk at the congress, he dwelled on the recent controversy over the Pfeiffer phenomenon—not to question its existence but to reinterpret it in light of the phagocyte theory. He ended his speech with a triumphant chord, "Concluding our overview, we must recognize the victory of the cellular theory of immunity and the defeat of the purely humoral approaches." After the talk, Metchnikoff reported modestly in a letter to Olga, "The audience was apparently pleased as they applauded mightily, but this was because I spoke loudly and in a lively manner after Buchner's soporific talk."

Years later, during the celebration of Metchnikoff's seventieth birthday, Roux poetically recalled his friend's animated speech in Budapest: "I can see you now at the Budapest Congress in 1894, disputing with your antagonists; with your fiery face, sparkling eyes, and disheveled hair, you looked like the demon of science, but your words, your irresistible arguments raised the applause of your audience."

What Roux failed to mention was that at that same congress, it was his own talk that had stolen the show. Behring was unable to attend, so it had fallen upon Roux to report on the first spectacular results obtained with the diphtheria serum in sick children. In the large trial he had undertaken in Paris in the preceding few months, the serum had cut mortality from diphtheria in half. The audience was overwhelmed. According to one American doctor who attended the congress, "hats were thrown to the ceiling, grave scientific men arose to their feet and shouted their applause in all languages of the civilized world. I had never seen and have never seen since such emotion displayed by an audience of scientific men."

24

KICKING AGAINST THE GOADS

AFTER THE BUDAPEST CONGRESS, THE year Metchnikoff turned fifty, a breeze of appeasement wafted through his high-ceilinged lab. The immunity war began to wind down, as did his combativeness. In June 1895, he published a paper in the *Annales* whose very title spelled reconciliation: "Extracellular Destruction of Bacteria in the Organism."

That same year, in September, Pasteur died after suffering another stroke. In the preceding months, Metchnikoff had visited him regularly. (Pasteur called him Monsieur Menshikoff, a more common Russian surname.) Metchnikoff invariably came away depressed from these visits. He might have already started worrying about his own aging. "It's tough to see this slow extinguishing," he once confessed in a letter to Olga after paying a weekly visit to Pasteur. In his grief over the passing of the great scientist who had done him so much good, Metchnikoff reminded himself that science had already lost Pasteur several years earlier.

In a sign of a let-up in his battles, in 1897 Metchnikoff took the time to write an article on a new theme that was to preoccupy him for the rest of his life: "A Review of Several Works on Senile Degeneration," published in *L'Année Biologique.* He had begun to fear aging and disease. The following year, when he started having problems with his kidneys at age fifty-three, he imagined he had a terminal illness.

The researcher who only recently had been swallowing deadly germs in the name of science now adopted a sterile diet, avoiding foods that might contain microbes. His preoccupation with his health verged on hypochondria. "Macabre thoughts pursued him," Olga wrote. He calmed down only after a kidney expert in Germany assured him he was out of danger.

As for Olga, she had never left Metchnikoff's side at the height of his battles, but now she spent only part of her time assisting him. In 1898, after ten years in France, the Metchnikoffs bought a house in Sèvres, a small town west of Paris, where they had been spending the summers for a number of years. In the backyard, they planned to build a studio for Olga. Almost forty, she had finally turned to her true calling. She started painting, mainly producing landscapes and portraits, soft and poetic like herself, including a pastel portrait of her husband. And she was taking sculpting classes with the famous Jean-Antoine Injalbert, who had just created four colossal bronze figures for the new Mirabeau Bridge in western Paris.

Even in peacetime, the work in Metchnikoff's lab was as hectic as ever. "I don't even know if there's an opera in Paris," he told a visitor from Russia, expressing what to him was the epitome of a sacrifice.

Just then, two final rounds of cannon fire shook up the phagocyte theory. Ironically, one was fired from within Metchnikoff's own camp.

He had let into his phagocyte fortress an exceptionally talented Trojan horse. Jules Bordet, a young, unassuming Belgian doctor with water-blue eyes and a Mona Lisa smile, had walked into his lab at the very beginning of the Pfeiffer crisis. Unlike Metchnikoff, he had no penchant for inventing grand theories. Rather, he was a consummate experimenter. He had long admired Metchnikoff from afar, but once in Metchnikoff's lab—right under the nose of his "dear and revered *maître*," as he called him in scientific papers—the starstruck Bordet quietly buried the phagocyte theory.

Bordet supplied the humoral theory with an entire series of critical concepts, removing major obstacles to its widespread acceptance. First of all, he revealed that the immune serum killed bacteria through the combined action of two substances: the antibody, discovered by Behring, and yet another substance, which ultimately came to be known by its German name, *complement*. The antibody provides a precise fit to the germs, whereas the complement, always present in blood, is a substance of a more general sort. This research later helped develop blood tests for syphilis and other diseases.

And if the discovery of the complement were not enough of a sacrilege in the shrine of phagocytosis, Bordet next revealed that immune fluids destroyed not only bacteria but all foreign material, such as red blood cells from another animal species. The finding crowned humoral immunity as a universal phenomenon, not limited to killing bacteria, and led to a forensic test for distinguishing animal blood from human at crime scenes.

It would have been only human if Metchnikoff were incensed with Bordet for conducting studies that played into the hands of the rival camp. Indeed, he treated students flirting with the humoral camp as infidels, once furiously scolding a new trainee who had naively published an article about humoral immunity in *La Presse Médicale*. But improbably enough, Metchnikoff never classed Bordet as a dissenter. They stayed in touch when Bordet returned to Belgium. And had Metchnikoff lived a few years longer, he would surely have been proud to see Bordet awarded the 1919 Nobel Prize in Physiology or Medicine "for his discoveries relating to immunity," most of them made in Metchnikoff's lab.

The key to their friendly relationship was that Bordet, whether by loyalty or conviction, was always careful to applaud the phagocyte theory, never crossing over to the enemy side in his interpretations of his findings. He probably further endeared himself to Metchnikoff by mounting an independent offensive against the last of Metchnikoff's formidable adversaries in Germany.

Paul Ehrlich was blond, slightly built, and ten years younger than Metchnikoff. Once dubbed "a brilliant eccentric," he puffed on strong Havanas throughout his waking hours, just as incessantly turning out epoch-making scientific ideas. Invisible uniting threads had run through his and Metchnikoff's backgrounds. Both had grandfathers who were secular Jewish intellectuals. In fact, had Metchnikoff's Polish-born grandfather Leiba Nevakhovich, instead of going east to Russia, followed the more common trend of fleeing westward to escape anti-Semitism, he might have very well found himself moving in the same circles as his contemporary Heymann Ehrlich, Paul Ehrlich's grandfather, in the neighboring Prussian province of Silesia.

As a young doctor in love with chemistry, Ehrlich had been so enamored of dyes that he used to draw laughs as he walked around with multicolor fingers. But it was Ehrlich who was to have the last laugh. He discovered a way to visualize tuberculosis germs better than even Koch had, creating widely accepted methods for staining various germs and cells. Yet another feat had turned him into a celebrity: producing the standardized diphtheria serum that had enabled Behring to forge ahead with his serum therapy.

A devout fan of Sherlock Holmes mysteries, Ehrlich next preoccupied himself with solving the mystery of immunity: explaining the activity of protective substances in the body fluids. And fatally for phagocytes, he managed to crack the case.

In the closing years of the nineteenth century, Ehrlich formulated his own theory of immunity, which explained how antibodies hooked up with toxins or germs. The secret, he said, was a tiny latch enabling the binding. This was the seminal notion of receptors, which Ehrlich called *side chains.* Antibodies could target specific germs thanks to a precise fit, just as a lock fits a key, a metaphor Ehrlich borrowed from organic chemistry.

The implications of the lock-and-key idea were momentous. They extended far beyond antibodies, or, for that matter, far beyond immunity. Ehrlich himself was later to apply the notion of receptors to the search for drugs, which he famously called "magic bullets," laying the foundations of the entire modern pharmaceutical industry. The Hollywood film about him made in 1940, two and a half decades after his death, was appropriately titled *Dr. Ehrlich's Magic Bullet.* Antibodies, too, could be regarded as magic bullets, manufactured by the body itself. Once defined by Ehrlich in this manner, they definitively steered immunology toward chemistry—that is, toward interaction between molecules—and away from cells.

Ehrlich's side chain theory might not have had as strong an impact had he not illustrated it with vivid pictures, each worth a thousand experiments. The technology was not yet available to visualize the receptors, but Ehrlich depicted them from his imagination. His drawings left no doubt in the viewer's mind the receptors were real—they resembled exotic plants or marine animals. Ehrlich himself likened his side chains to "grasping arms" that operated like the tentacles of sundew, a carnivorous plant. He drew these memorable pictures everywhere—on the doors and walls of his office, on linen tablecloths when invited to dinner, even on a listener's cuff. On one occasion, when explaining the side chain theory to Koch, he grabbed some chalk and covered the floor with receptor-antibody cartoons, rolling up the carpet as he ran out of space.

Most scientists enthusiastically welcomed the side chain theory, but not so Jules Bordet, who took issue with the mechanisms Ehrlich defined. He was particularly bothered by Ehrlich's drawings, calling these "puerile graphic representations" unscientific.

Alas, Bordet ended up playing into the hands of the humoral camp one more time by keeping the combative Ehrlich on his toes. Ehrlich had a large file labeled "Polemics," with outlined strategies for each of his ongoing disputes. He labeled a few series of experiments *gegen Bordet*, "against Bordet." Others were directed against Metchnikoff, Roux, and other Pasteur Institute scientists. Every day, Ehrlich dispatched to assistants his famous *Blöke*, the colored index cards, with army-like instructions: "I think that the Pasteur people are preparing a similar attack and I'd like to beat them to the punch"; "I was just thinking how much has to be done in the next two short weeks: 1—anti-Metchnikoff, 2—anti-Besredka" (Alexandre Besredka was a doctor in Metchnikoff's lab); "More important is that we continue with all energy the anti-Metchnikoff matter"; and "The main thing is to finish the anti-Bordet work."

Metchnikoff, unlike Bordet, didn't seek to disprove the side chain theory, only to try and deal with it. "I witnessed the concern and anxiety with which he received each new publication having to do with the role of antibodies in blood," reported a Russian doctor who trained with Metchnikoff in Paris at the end of the nineteenth century. "Ehrlich's humoral theory of immunity, emerging just then, represented in his eyes a tremendous threat to the integrity of his phagocyte theory as the basis of immunity." That same Russian trainee further testified that all new findings from the humoral camp "were immediately thoroughly checked in Ilya Ilyich's lab, and presented in the new light of Metchnikoff's cellular immunity, in which phagocytes preserved their monopoly in one way or another."

New insights into immunity poured from these studies. Metchnikoff figured out with dyes, for instance, that acids sometimes help phagocytes digest their prey, similarly to the way acidity contributes to digestion that takes place in the stomach. (He had no tools to fully disclose the cells' actual killing mechanism, a complex process that continues to be investigated today.) He also showed that invertebrates don't make antibodies. Indeed, these immune weapons appear only higher on the evolutionary ladder, in vertebrate animals. And he continued to insist that the antibodies were made by cells. But one central question remained unresolved: which cells were these?

Ehrlich argued that antibodies were manufactured by the body organ under attack. He insisted, for instance, that because tetanus affected the nervous system, the antitetanus antibodies were made by nerve cells. Metchnikoff was determined to prove him wrong. Since the nervous system could not be temporarily obliterated so as to test this point, he decided to experiment with an arguably dispensable body part. He injected male rabbits with "spermotoxin," a substance believed to have a toxic effect on sperm. By Ehrlich's reasoning, the rabbits were expected to manufacture antibodies in the affected cells, in this case, sperm cells. Metchnikoff, however, proved that the castrated, spermless rabbits were just as successful at making the antibodies as were their intact counterparts. Writing in the *Annales,* he concluded, "Antispermotoxin is not produced by male sex cells, but by other cells, whose nature at present can be only postulated."

It is not difficult to guess which "other cells" he had in mind. He was wrong—and so was Ehrlich. We know today it is neither phagocytes nor the cells in attacked body organs but small, round white blood cells called *B lymphocytes* that make antibodies. Metchnikoff ignored the lymphocytes, as did most other scientists at the time. Compared to the larger, voracious phagocytes, they must have looked boring to him. Besides, invertebrates, which had claimed so much of his attention, have no lymphocytes at all.

Metchnikoff around the time he graduated from high school in 1862.
© Archive of the Russian Academy of Sciences (584.2.245)

Metchnikoff and Olga in the late 1870s or early 1880s.
© Archive of the Russian Academy of Sciences (584.2.254)

Olga, Metchnikoff (seated second from right), and his Russian trainees at the Pasteur Institute, 1890. Standing on the left is Waldemar Haffkine.
© Archive of the Russian Academy of Sciences (584.2.261)

Metchnikoff and Émile Roux.

Metchnikoff, Olga (right), and Marie Rémy.

Metchnikoff and Lili, around 1905.
© Archive of the Russian Academy of Sciences (584.2.322)

"Manufacture of Centenarians": Cartoon of Metchnikoff by B. Moloch in *Chanteclair*, 1908, No. 4, p. 7.

Metchnikoff and Leo Tolstoy in Yasnaya Polyana, May 1909.

Metchnikoff and Olga during their trip to Russia in 1909.
© Saada Collection

Metchnikoff in his Pasteur Institute laboratory, 1914.

Drawings from Metchnikoff's *Immunity in Infective Diseases*:

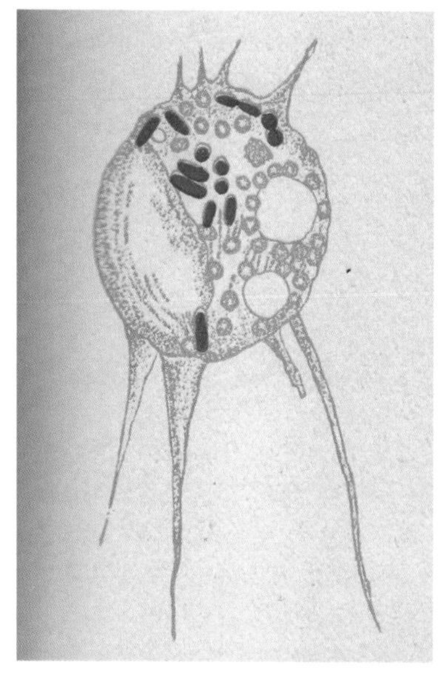

Guinea pig macrophage
with *E coli* bacteria

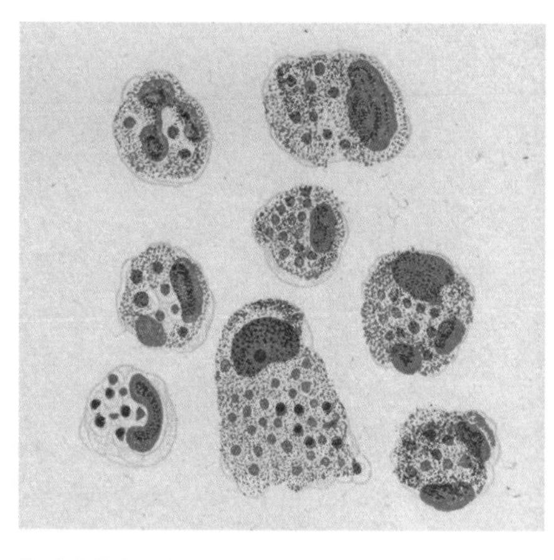

Rabbit leukocytes with tetanus spores

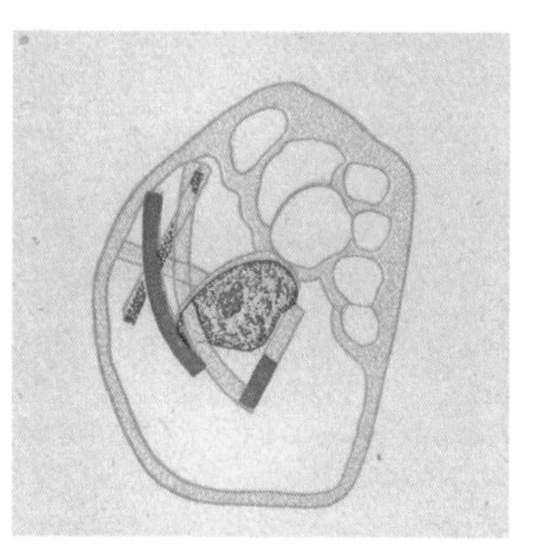

Rat liver macrophage
with bacteria

Metchnikoff's new imperative became to make peace with his ultimate nemesis. He started arguing that his and Ehrlich's theories simply operated at different levels. He wrote, "Ehrlich's theory does not at all contradict [the phagocyte theory]; it only strives to penetrate deeper into the mechanism of interactions between the cell and the microbe." He was absolutely right. Receptors operate at the molecular level, on a scale much smaller than that of cells; if a cell were the size of the Earth, a receptor on its surface would be roughly the size of Manhattan.

But it was too late. The two immunity camps had diverged too far in their respective directions. Besides, Metchnikoff and Ehrlich prayed in different temples. Ehrlich's religion was chemistry. Metchnikoff's worship of evolution was foreign to him. Metchnikoff was the one to make all the efforts at reconciling their theories, probably fearing to lose his last duel.

For all their scientific disputes, he and Ehrlich maintained a courteous relationship. They had been meeting at medical congresses for years, starting probably with the 1891 London Congress they both attended. Ehrlich, who deluged listeners with impassioned discourse on his research and had been dubbed Dr. Fantasy by an opponent, was one of the few people who could out-talk and out-fantasize Metchnikoff. Their encounters could even be described as effervescent. "Both exceptionally lively, they picked up on each other's thoughts on the fly," recalled one witness.

Ehrlich acknowledged Metchnikoff's charisma and seems to have enjoyed hobnobbing with him, profusely thanking him for a cordial welcome after each trip to Paris. In correspondence with friends and colleagues, however, he vented his true sentiments without any inhibitions. Never missing an opportunity to use a Latin phrase, in 1899 he wrote to a friend in Denmark that Metchnikoff was the "real *spiritus rector*" (guiding spirit) behind the attacks from Paris and that "with his captivating personality, he had Roux completely entrapped." Ehrlich added further that it was "an awful shame that such an experimenter and such a clear head as Roux should have strayed so deeply into mysticism and Russian fog!"

Metchnikoff, he continued, will spare no effort to "choke my newest ideas to death because they attack his phagocyte theory at the jugular." At around the same time, in a letter to another Danish colleague,

Ehrlich stated that once certain study results would be published, Metchnikoff "will find it difficult to kick against the goads, so that I might soon be able to switch to a frontal attack." Writing to a colleague in Frankfurt in 1899, he announced, alluding to Potemkin villages, that in the wake of one of his studies, "entire Metchnikoffian villages and palaces will tumble down."

When delivering the prestigious Croonian lecture in London in 1900, Ehrlich inflicted upon Metchnikoff the ultimate insult, dismissing his ideas in a single sentence in a cruel past tense, as if his years of hard work had been nothing but a nice try. In a historical overview he gave before describing his side chain theory, he remarked, "The very able investigations of Metchnikoff and his theory of phagocytosis were, to many investigators, inconclusive."

25

A ROMANTIC CHAPTER

THE YEAR 1900 CAUGHT METCHNIKOFF at a brink. Like the year itself, which teetered between closing one century and opening another, he was torn between two battlegrounds, the old and the new: defense of his theory of immunity, or pursuit of new frontiers for his phagocytes.

France too was being pulled in different directions. The spirit of the period the French had somberly called *fin de siècle* still lingered, but the beginning century brimmed with hope, new artistic movements, and technological innovation. People already drove cars, were entertained by motion pictures, and soon would pilot airplanes. In politics, French society was torn apart by discord, later to be dwarfed by the conflagrations the century had in store. Royalists still sought ways to foil the Republic; socialists urged the workers to hold as many strikes as possible; anti-Dreyfusards had been bolstered by a recent upsurge of nationalism. In the 1900 municipal elections in Paris, nationalists gained a majority for the first time, foreshadowing a future upswing of xenophobia.

In the spring of that year, political divisions were set aside for a short while when the gigantic Exposition Universelle opened on both banks of the Seine, showcasing the magic of *fée électricité* via a moving sidewalk, thousands of incandescent lamps illuminating monuments, and the first line of the Paris Metro—launched for the occasion despite doomsday warnings of dangers that at first earned it the nickname the "Necropolitain." Over the span of six months, more than fifty million people visited the exposition, among them nineteen-year-old Pablo Picasso, who showed a canvas in the Spanish pavilion. In the newly erected glass-vaulted Grand Palais on the Champs-Elysées, thousands of works were displayed in the main show of modern art. The Impressionists were well represented, even though they were still more popular abroad than in France. In the sculpture halls, there were so many equestrian figures that one journalist quipped he had the impression of being at a horse show. The 500 or so prizes awarded in the sculpture category included nearly 150 bronze medals; one of these went to Olga Metchnikoff, who exhibited two busts—a dervish and a woman.

This would remain her highest honor as an artist, and it signaled a switch. Olga was now to devote herself primarily to painting and sculpting, though she did continue to assist Metchnikoff in the lab when needed and translated his writings from French into Russian. Her husband's phagocyte theory was eighteen years old, an age at which it could be expected to manage on its own.

Olga applied herself to aiding Russian artists—organizing, together with friends, a society called Montparnasse that helped visiting Russians to pursue art interests in Paris. Wearing long, plain dresses, often in her favorite grayish-lilac color, and antique jewelry, she would over the years engage in numerous charitable activities in support of Russian émigrés in Paris. She donated her works to an art auction for the benefit of needy students and took part in the work of the "Hungry Friday" shelter.

On the rainy morning of August 3, while Olga was still awaiting the announcement of the exposition's prizes later that month, Metchnikoff waged his final immunity battle, an aging boxer dignified in his resolve to give his opponents one last fight. There could hardly have been a more reassuring setting for his grand finale. Not only was the Thirteenth International Medical Congress held in Paris, the city Metchnikoff called home, but the Bacteriology Section convened within the comforting walls of the Pasteur Institute, in its large new amphitheater, with plain white walls, rows of long wooden benches, and shaded electric lamps hanging low from the ceiling. With his usual passion—"He was like Sybil on a tripod," a colleague had once described his delivery at a conference, referring to the inspired prophetess of antiquity—he reported on his antispermotoxin studies, finally declaring, "We arrive at the conclusion that the humoral properties represent only a certain fraction of the ensemble of immunity, which is dominated by the cellular properties."

Roux, nursing his health in the country, cheered him on. "You have won a smashing victory, it seems that your opponents have lost their former belligerence. Peace—Pax Metchnikoviana—will now reign in the question of immunity," he wrote in a letter to Metchnikoff, alluding to Pax Romana under the influence of a novel he was reading about ancient Rome. "I rejoice with all my heart, since no scientist has ever fought [for his views] with greater force or courage."

The congress left Metchnikoff with a disturbing, and probably justified, feeling that his theory was not being properly understood. His zoological concept of cells changing their identity in the course of evolution was foreign to most medical researchers. Indefatigable in his naïveté, he still believed universal acceptance was around the corner, if only other scientists learned more about his phagocytes. So as soon as the rattling of the carriages carrying congress members away from the Pasteur Institute had faded on rue Dutot, he sat down to write his magnum opus on immunity.

The six-hundred-page *Immunity in Infective Diseases*, published in France in October 1901, is a gargantuan ode to phagocytes. Metchnikoff dedicated it to his *chers amis*, Roux and Émile Duclaux, the gentlemanly chemist who had succeeded Pasteur as the institute's director, both of whom had initially rejected his theory—with the implicit hope of similarly converting every reader from foe to friend. Immediately translated into German, then into Russian and English, the book is a noble attempt at securing a Pax Metchnikoviana with other theories, even if it does proclaim that the phagocyte theory can explain all aspects of immunity.

Immunity in Infective Diseases breathes with Metchnikoff's desire to be of service to humanity. "Science is far from having said its last word, but the advances already made are amply sufficient to dispel pessimism provoked by the fear of diseases and the feeling that we are powerless in our struggle against them," Metchnikoff wrote in the book's closing sentence. "[The book] gives one the definite impression that for Metchnikoff the phagocyte theory was more than a favorite brain child, to be defended at all cost," science historian Gert Brieger wrote decades later, in a preface to a reprint edition. "The theory's implications were in the direction of an amelioration of the human condition."

Immediately upon publication the book received glowing reviews. The *British Medical Journal* went overboard, noting that it "reads as pleasantly as a good novel; some, indeed, will think far more pleasantly." Praise came even from unexpected quarters. Metchnikoff reported in a letter to Bordet: "Your favorable attitude was not unexpected. I was much more surprised to receive a very nice letter about my *Immunity* from Mr. Ehrlich."

It was all to no avail. For all the applause, the book made few converts. Most damaging to Metchnikoff was the fact that it stirred very few objections. His opponents, exhausted by nearly two decades of confrontations, had stopped fighting him. And once deprived of resistance, he lost steam. Constant opposition may be difficult, but the worst blow is when the enemy deserts you.

After *Immunity*, Metchnikoff abandoned immunity.

About two months after the book was published, in December 1901, the Nobel Prizes were awarded in Stockholm for the first time. The prize in medicine was presented to Emil Behring, for having "placed in the hands of the physician a victorious weapon against illness and deaths." As was to be expected, Metchnikoff immediately wrote to his friend "to express my heartiest and most sincere congratulations upon this high honor, which you so well deserve." The close timing of his book, which charted future horizons for knowledge, and of the prize, awarded for a concrete achievement, embodied Metchnikoff's abysmal isolation at the closing of the immunity war, a lone voice prophesying far-flung visions in a world bent on rewarding practical successes.

By then, Metchnikoff had assured his place in the history of immunology. He was widely recognized for having shaped the modern concept of immunity, the very notion that the human body possesses inner powers to resist infection. "To Metchnikoff, we owe the first serious attempt to explain the resistance of the body to bacterial invasion," the *Lancet* was to write in his obituary. "He formulated the conception of an active bodily defense." Besides, his contests with the various German scientists provided a fertile ground for immunity research to advance by leaps and bounds. When Cambridge University Press published a major volume on immunity in 1904, it was dedicated "to Paul Ehrlich and Elie Metchnikoff, whose genius and influence have so greatly advanced and stimulated the search after truth amid the complex problems of immunity." By the time his *Immunity in Infective Diseases* appeared, Metchnikoff had already been made *officier* of the Legion of Honor by the president of France and elected to the Royal Society, the American Academy of Arts and Sciences, and France's Academy of Medicine, among numerous other honors.

Still, the outcome of Metchnikoff's struggles over phagocytosis lent itself to strikingly different interpretations. His self-proclaimed admirer Joseph Lister declared fondly in a presidential address to the British Association for the Advancement of Science, "If ever there was a romantic chapter in pathology, it has surely been that of the story of phagocytosis." Lister didn't elaborate, but Metchnikoff's championing of phagocytosis could surely be seen as romantic—in its pioneering spirit, its reliance on the power of imagination, and its vision of human beings as an indelible part of nature. In contrast, an Odessa physician who had fallen out with Metchnikoff suggested disdainfully in a bacteriology textbook that Metchnikoff had lost his battles. "Phagocytosis has been a story of disappointment," he wrote.

In fact, as is the case in most serious scientific debates, there were no losers in the immunity war. But even if many researchers by now believed phagocytosis did play a role in immunity, this role was thought to be secondary. Besides, the study of phagocytes was of no help at all in treating disease, not for another few decades. Metchnikoff's greatest pride, that his theory harked back to ancient animals, had turned out to be its weakest point. Who cared how the comb jellies defended themselves against microbes? Far more burning issues were at hand. Hospitals still overflowed with tuberculosis patients, cholera still disrupted world travel, and typhoid fever was still deadlier to armies than enemy fire. The solutions were thought to lie submerged in the body fluids.

So for the time being, humoral immunity seemed to have triumphed over cellular. The verdict was handed down by scientists in their labs—they cast votes with their microscopes. Only Metchnikoff's disciples at the Pasteur Institute still studied phagocytes, and a handful of British physicians tried to reconcile the two immunity camps. Other than that, it would be more than half a century until scientists elsewhere were to pick up his phagocytosis research.

Metchnikoff himself was as convinced as ever in the truth of his own vision of immunity. Luckily for him, whatever bitterness he might have felt was drowned in the flurry of new studies in his lab. He was already totally immersed in pursuing a new goal, no less than to defeat aging.

IV

NOT BY YOGURT ALONE

26

HAUNTED

IN PHOTOGRAPHS OF HIS LATER years, Metchnikoff appears as defiant as ever. Curiously, by his midfifties, with his shaggy gray hair and beard, he had come to look every bit an aging nihilist. Though in his youth he had never engaged in outward displays of protest, he now finally caught up with the rebellious look, appearing messy enough to pass nihilist muster. Journalists delighted in describing his "undisciplined hair, lying in heaps, tangled like wheat after a thunderstorm," as well as his "big head, which seems expressly fitted to butt against obstacles," his "small, mischievous eyes of extraordinary vivacity," and "the guttural, flexuous, lilting voice that gives his speech its charm."

His girth had been expanding and his gait had grown heavier. Arriving in the lab, he took off his worn felt hat (on which he occasionally sat when excited), put aside his umbrella, and removed his galoshes, which he wore year-round in damp weather. The huge pockets of his shabby suits always bulged with books, newspapers, and letters, so that the Paris correspondent of the *Los Angeles Times* once wrote that he looked "much like one of those Bohemian book hunters or print collectors who haunt the quays of the Seine." He often began his sentences with a favorite expression, *Donc alors,* which could roughly be translated as "Well . . ."

Students and others were now calling him Père Metchnikoff—that is, Papa Metchnikoff—or simply Papa Metch. From a young man nicknamed *Mamasha* he had morphed into a buoyant, obsessively caring paterfamilias. He had a way of making people feel they were obliging him if they turned to him for assistance. He regularly loaned money to students, treating them with fatherly attention. "He was more than a father to me," a scientist who in his early career worked in Metchnikoff's lab recalled years later.

That lab was open to everyone. Old and young came to Papa Metch for medical advice, even though he was not a physician. He had wanted to become a doctor in his youth, before his mother had talked him out of it. Now this early calling entered his life through the side door. Metchnikoff took care of the medical needs of dozens of people, among them total strangers, subjecting them to tests, finding the best doctors to treat them, and serving as an interpreter if they spoke no French.

Not necessarily waiting for people to turn to him, he stepped in himself whenever he could. Noticing that a newspaper vendor on the Montparnasse train station had a persistent cough, he personally tested her mucus for tuberculosis. He frequented a watchmaker's shop near that same station, and when he learned that the watchmaker's wife was gravely ill he asked to examine her and urgently summoned a surgeon, who saved the woman's life after diagnosing her with appendicitis and operating on her. Metchnikoff acted here with the same sense of mission with which he engaged in his research, a legacy from his youthful imperative to be of service to the people. "Since I'm paying such attention to the sick from the higher echelons, my conscience doesn't allow me to neglect the lower fellows," he once wrote to Olga, explaining why, after checking up on the health of an established artist, he made a trip outside of Paris to examine a friend's concierge who was ill with kidney disease.

An appeased, middle-aged Metchnikoff was much happier than the volatile quarreler he had been for years. "The period between fifty and sixty-five years of age was the happiest of his life," Olga wrote. But the contentment had come at a price. "The intensity of his feelings and sensations had greatly decreased since he'd reached a mature age," Metchnikoff wrote in a veiled autobiographical passage in one of his books describing the blissful indifference of a scientist "friend." He no longer suffered from sharp noises but enjoyed music less, savoring silence; he developed a taste for such plain food as bread and water. And he was not as easily hurt as he had been before. "Consequently, his character became much more pleasant for others and incomparably more balanced," Metchnikoff wrote about himself in the third person with remarkable self-awareness.

There was one dark spot: anxiety about death gnawed at him. It had begun to vex him in his early fifties, at the time his kidneys had started giving him trouble and he felt his health had started to falter. "Observing in his organism certain signs of illness that could turn lethal, he experienced an entirely peculiar sensation: an instinctive fear of death," he confessed in the veiled autobiography. Judging from the frequency with which he returned to the topic in his lectures, articles, and books, this ever-present fear haunted him. As a lifelong atheist, he was denied the comfort of believing in an afterlife. So as always, he turned to science for solutions.

The immunity war behind him, a mellowed Metchnikoff took on new enemies that were no less formidable than his former German opponents: aging and death. Still an idealist at heart, he strived to free not only himself but all of humanity of these two evils.

He was surely aware that over the centuries, attempts to delay or reverse aging, and to extend life, had been as innumerable and varied as were theories of aging. In ancient Rome, old men and women rushed into the arenas to try to rejuvenate themselves by drinking the blood of dying gladiators. Blood of young animals or people had been traditionally viewed as a conduit of vitality. Much later, in the second half of the seventeenth century, the physician to King Louis XIV transfused the blood of a lamb into a patient. Such transfusions were attempted for rejuvenation—with disastrous results. In a bon mot of the critics, the procedure required three lambs: one to serve as the donor, one to be the recipient, and one to perform the procedure.

A lower-tech method, known as *gerokomia*, had been practiced with medical approval from the time of King David to modernity. "In biblical times, it was thought that contact with young girls rejuvenates enfeebled old men and prolongs life," Metchnikoff wrote about *gerokomia* in one of his books. "It found a following even in recent times." Cristoph Hufeland, physician to the king of Prussia and author of the doctrine of macrobiotics, wrote in his 1796 bestseller, *Art of Prolonging Life*, "The breath of young girls contains the vital principle in all its purity."

For most of history, a common belief behind the search for immortality or eternal youth was that humans had been originally designed to live forever but were stripped of this privilege because they had sinned. After being expelled from the Garden of Eden, Adam still managed to live to 930 years, and his descendent Methuselah, the biblical champion of longevity, to 969. Noah was 600 at the time of the Flood. But sins must have accumulated over generations, for Abraham lived just to 175. For those reading Genesis literally, as most educated people did in the seventeenth century, it was clear that prolonging life meant reclaiming a longevity that had somehow been lost.

All that changed in the eighteenth century; the quest to prolong life had entered a stage that fit in with Metchnikoff's belief in the power of science. As Gerald J. Gruman argues in *A History of Ideas about the Prolongation of Life*, during the Age of Enlightenment, with the rise of the idea of progress, people started thinking about life extension as a future goal, rather than a long-lost ideal. Humans were conquering nature, so why wouldn't they one day triumph over death? Benjamin Franklin, for one, ventured the prediction that life expectancy might one day exceed that of the pre-Flood patriarchs. "The rapid progress true science now makes, occasions my regretting sometimes that I was born so soon," he wrote to a friend in 1780. "It is impossible to imagine the height to which may be carried, in a thousand years, the power of man over matter. . . . All diseases may by sure means be prevented or cured, not excepting even that of old age, and our lives lengthened at pleasure even beyond the antediluvian standard."

It was with scientific progress in mind that the great eighteenth-century Scottish anatomist John Hunter dissected thousands of corpses and countless animals, many of them living. In one of his lectures on surgery, he dwelt on a chilling idea for increasing human longevity: "I had imagined that it might be possible to prolong life to any period by freezing a person in the frigid zone, as I thought all action and waste would cease until the body was thawed. I thought that if a man would give up the last ten years of his life to this kind of alternate oblivion and action, it might be prolonged to a thousand years: and by getting himself thawed every hundred years, he might learn what had happened during his frozen condition." Hunter tried out his idea on carp, freezing then thawing them very slowly. To his utmost disappointment, the carp failed to revive.

In June 1889, when Metchnikoff was busy equipping his new lab at the Pasteur Institute, noted neurologist Charles-Édouard Brown-Séquard, then seventy-two, unveiled a rather extravagant recipe for reversing aging—astounding members of France's Biological Society, of which he was president, with a report on his own rejuvenation. Having ground up testicles of young guinea pigs and dogs and injected himself with the extracts, he claimed to feel invigorated intellectually, adding that his "other forces" too had been notably improved. Alas, his "elixir of life," as the press called it, at first seized upon in different countries by thousands of physicians, not to mention scores of charlatans, failed to deliver on its promise. Nor did it invigorate Brown-Séquard himself for too long: he died of stroke at seventy-seven. His outrageous experiments did stimulate research into internal secretions, eventually leading to hormone replacement therapy, but because the rejuvenation brouhaha around them proved so misguided, they had little immediate effect on the study of aging.

Thus, despite the mass of scientific and not-so-scientific attempts to delay old age, solid scientific knowledge about aging was almost entirely lacking at the time Metchnikoff directed his energies to longevity research. Several landmark treatises on diseases of the elderly had been published in the late nineteenth century, but the causes of aging itself—and, for that matter, the causes of our limited life span—remained a mystery.

Few people had survived beyond the ages of seventy or eighty years, even fewer beyond one hundred. A hidden biological law seemed to dictate the length of life span. When it came to aging, a central quandary loomed particularly large: is it a normal life process, just as normal as, say, puberty—or is it an avoidable disease? "It's really surprising that so few precise facts have been gathered in a biological area of such great importance," Metchnikoff had written in his review on aging, the one he submitted in 1897 to *L'Année Biologique.* "We think science has a great deal to gain from applying itself to the study of this matter."

He was about to become, yet again, a founder of a major new area of research: systematic investigations into the biology of aging. It's tempting to wonder whether more than a hundred years ago, breaking new ground was easier than it is today because there was more ground to break. The truth, of course, is that the unknown is always vast. At any given time, it yields its secrets only to intrepid minds.

27

VIVE LA VIE!

One day in late December 1899, a few months before his last immunity battle at the Paris Congress, Metchnikoff had woken up to discover he had found the key to immortality. That, at least, was what the popular French daily *Le Matin* announced on its front page that morning. A jubilant headline blared in block letters: *VIVE LA VIE!* [LONG LIVE LIFE!]—Doctor Metchnikoff and the Struggle Against Death. Subheadings throughout the article, which dwelt on Metchnikoff's new research on aging at the Pasteur Institute, were no less dramatic: A Visit to the Alchemists on Rue Dutot—The Elixir of Eternal Youth—At the Institute of Miracles—Old Age Defeated. A few days later, in an editorial on the prospects for the upcoming century, the paper gushed, "None of us should despair to see the year 2000! We'll reach the age of the patriarchs, and Monsieur Metchnikoff will be damned only by heirs to fortunes."

Before long, press around the world picked up news of the miracles being wrought on rue Dutot. The horrified Metchnikoff started receiving letters from the elderly in France and elsewhere begging him to help them not to die. "Metchnikoff is really annoyed by the noise journalists make around his name," Jules Bordet wrote to his wife in Belgium. "It's a bit of his own fault, he should have chucked them out more vigorously."

But chucking people out was out of character for Metchnikoff. Journalists, his comrades-in-arms in the sacred mission of popularizing science, had always appreciated the ease with which they could knock on Metchnikoff's door at any time, be it to discuss his own work or that of scientists elsewhere. So when the sensational headlines began to appear, Metchnikoff countered the reports by granting ever more interviews. In February 1900, he pleaded with a *Los Angeles Times* reporter, "I've been a victim of the hasty journalists. I work and hope, but I promise nothing. I regret very much—please say that I regret very much—all the talk that has arisen about my researches."

In his research on aging, he had been focusing his attention on phagocytes. He knew that in addition to defending the organism, these cells performed maintenance, devouring damaged tissue. In studying autopsy samples of the elderly, he came to the conclusion that the phagocytes had devoured a portion of their brain and body tissues.

Remarkably, after the phagocytes were done devouring, the results appeared the same in damaged organs and in aged ones. When Metchnikoff looked, for instance, at defective kidney tissue under the microscope, he was unable to tell whether the defects had been produced by microbes, alcoholism, chronic lead poisoning, or old age. In all these cases, some of the healthy kidney cells had been replaced with scar tissue. Similarly, in the bones of the elderly, healthy matrix had been eaten up by giant cells, leading to osteoporosis. Exactly the same happened when bone was invaded by microbes causing leprosy or tuberculosis.

Metchnikoff wondered: *Was aging simply a chronic disease caused by microbes or poisons?* His answer: "Aging is a disease that should be treated like any other."

By then, he was too much a prisoner of his own grand discovery to see anything but phagocytosis when he looked through his microscope. He hypothesized that macrophages or other phagocytes, provoked by infection or poisoning, were responsible for the deterioration of tissue in old age. In other words, as in his theory of immunity, his beloved phagocytes were yet again granted a star status—but surprisingly, they were no longer heroes but villains. Aging appeared to him as a Darwinian struggle of sorts between stronger and weaker cells. "Senile degeneration can be reduced to macrophage activity, which leads to the destruction of the noble elements incapable of defending themselves," he wrote.

We now know that aging is an extremely complex process, nudged along by hundreds of genes. Phagocytes are but one theme in this symphony. Their primary job is to maintain tissues in good order, even if their accumulated effects lead to damage later in life. Yet Metchnikoff, certain he had identified the villains, started developing strategies to protect body tissues against phagocytes. Perhaps a vaccine could stimulate "noble" tissues to counter the phagocytes? Or maybe a vaccine could eliminate the microbes that overly excited the phagocytes? He tried, unsuccessfully, to prepare the latter vaccine by extracting antimicrobial substances from such startling sources as digestive juices of moth larvae, flies, and dung beetles, whose innards are known to destroy microbes.

While in the midst of these studies, Metchnikoff had been careless enough to discuss his ideas with journalists. That was how the headlines about the elixir of youth had come about. Hadn't the House of Pasteur already conquered anthrax, rabies, and diphtheria? Aging could have very well been next in line. The journalists trumpeted the report on Metchnikoff's ongoing research as yet another miracle in the making.

From then on, they were to hang on his every word, hailing him as "the successful Ponce de León," referring to the sixteenth-century Spanish explorer who had searched for an island with a fountain of youth, instead finding Florida. And Metchnikoff, who turned into what today would be called a media personality, never disappointed them or their readers, promising at least to reveal the mechanisms of aging if not eliminate it altogether. When he took the time off to write *Immunity in Infective Diseases* after the 1900 Paris Congress, he couldn't wait to get back to the lab. He wrote to Olga, "In the free moments, my thoughts already turn to questions of atrophy and aging."

"Scientists have been neglecting aging, which nonetheless presents a problem of great interest," he announced in his first research paper on the biology of aging in the *Annales* in 1901. In that paper, he turned to the most telltale sign of aging: the whitening of hair. He drew research material from his own disheveled mane (as always, using self-observations for his studies, he periodically cut samples of his own hair in different stages of graying), from the fur of an old Great Dane, and from the beard of Professor Jean-Baptiste Charcot, then collaborating on a research project with his lab—son of the renowned neurologist with whom the young Sigmund Freud had trained just before Metchnikoff had moved to Paris.

Finding pigment-filled cells in all the graying hairs, Metchnikoff reached the conclusion that a subclass of phagocytes was causing the hair to lose color by gobbling up its pigment, and that the mysterious cases of sudden hair whitening could be explained by increased hair-phagocyte activity. (According to modern science, the whitening of hair is an extremely complex, genetically controlled process that has to do with the hair follicles ceasing to produce pigment. But Metchnikoff was not entirely off the mark in his interpretation, as phagocytes do play a role in removing pigment during the normal hair growth cycle.)

The newspapers had a great deal of fun with these findings. *Le Temps* evoked the famous case of Queen Marie Antoinette, whose hair had supposedly turned white overnight before she was beheaded. *Le Figaro* quoted Metchnikoff as saying that women who curled their hair with a hot iron were possibly delaying its whitening, since heat could damage pigment-eating cells. *Le Petit Parisien* cited a prominent Parisian physician as suggesting that the recently discovered X-rays might be even more helpful than heat in protecting the hair from the "color-eating cells."

Metchnikoff, in the meantime, was trying to figure out why we die. He soon developed a theory of aging aimed at extending human life.

28

LAW OF LONGEVITY

In April 1901, after crossing an unusually calm English Channel, Metchnikoff for the first time exposed his newly formulated theory of aging to the public in the notoriously rainy Manchester. He traveled there to receive the Wilde Medal of the Manchester Literary and Philosophical Society, the first foreigner to achieve this honor. In the society's compact lecture hall, hung for the occasion with scientific drawings by Olga, he delivered an hour-long lecture in French, "The Flora of the Human Body," in which he outlined his brand-new explanation of why we age too fast and die too soon.

As suggested by the title of his lecture, the culprit, he announced, was the body's flora—microscopic organisms inhabiting our internal organs, primarily the large intestine, or colon, the body's largest microbe container. The idea that waste products in the intestines poison the human body went back at least to ancient Egyptians. In the late nineteenth century, with the establishment of the link between germs and disease, this belief had gained new validity, turning into a short-lived obsession among physicians. The contents of the gut were thought to putrefy and release toxins through the action of bacteria. As Metchnikoff noted in his lecture, some fifteen years earlier French pathologist Charles-Jacques Bouchard had adopted the term *autointoxication,* or self-poisoning, to describe this notion. Physicians were attributing anything from headaches and fatigue to heart disease and epilepsy to autointoxication, having their patients swallow disinfecting mixtures containing charcoal, iodine, mercury, or naphthalene to "sterilize" the intestines. (Autointoxication fell into disrepute in the 1920s, but today it continues to be promoted in certain forms of alternative medicine and resurfaces in different guises in mainstream medicine.)

Metchnikoff conceded that intestinal flora could be beneficial too, referring to his own cholera studies, which had led him to suspect that in some people, gut microbes prevented the disease from erupting. But most of these microbes, he argued, exert a harmful effect on the body, "and this leads to premature aging of our tissues and organs."

Lashing out with a bitter invective against the colon, Metchnikoff, as a zoologist and a Darwinist, pointed to the animal origins of human beings. Humans can blame their tendency toward autointoxication on the slow pace of evolution, he said. In our evolutionary past, the colon had helped mammals to survive. It contained not only microbes that facilitated the digestion of plant food but also remnants of digested food, enabling the animals to chase prey and escape predators without stopping to empty their bowels. Birds, he pointed out, discharge their excrements in midflight, consequently having no need for a colon. Humans, on the other hand, he said, "derive no benefit from this organ," particularly since they cook their food, making it easier to absorb. Though the colon was already known to play a role in the absorption of water and minerals, Metchnikoff believed it was less essential in this respect than the stomach or the small intestine. He was certain the colon should have long been eliminated by natural selection, if only the latter were more effective.

What crystallized from these ideas was Metchnikoff's very own law of longevity. He took on the riddle that some of the greatest minds had tried to solve since antiquity: is there some hidden natural clock that dictates the length of life? Aristotle had noted that larger animals tended to live longer than smaller ones, an observation that works as a broad principle but has too many deviations to qualify as a general rule. In the nineteenth century, scientists had proposed a rule by which maximum longevity of living beings was said to be proportional to their period of growth. One French biologist pronounced this ratio to be five to one in mammals, which allotted humans a life span of about a hundred years. But the ratio was too arbitrary to furnish an explanation; a particularly upbeat version was seven to one, instantly extending maximum longevity to 140 years.

Metchnikoff proposed his own explanation. "I'm almost inclined to derive a general rule: the longer the large intestine, the shorter the life," he declared. Of the small animals that can be kept as pets, he said, birds such as canaries live up to twenty years, as do goldfish, and turtles live even longer. In contrast, mice live three or four years at most, and they are the only small pets to have a proper colon. The numerous exceptions to this rule did not disturb him in the least. He argued, for instance, that elephants lived long despite having a very large colon thanks to peculiarities of their intestinal flora.

During the festive seven-course dinner the evening after the lecture, "the Englishmen delivered long speeches showing their sense of humor and provoking laughter. I laughed with everyone else though I understood very little of what was being said," Metchnikoff reported to Olga. And just as he had promised his enraptured Manchester audience, upon returning to Paris he threw himself into the conquest of aging with the same abandon with which he had once defended his phagocyte theory, and with the same monomaniacal focus on his own beliefs to the exclusion of all contradictory evidence.

His lab was gradually filling with old animals: mice, rats, geese, a graying eighteen-year-old dog, and an unusually perky twenty-three-year-old cat. He reserved a place of honor for parrots, the object of his second *Annales* paper on the biology of aging. Metchnikoff noted that certain parrot species live almost as long as do humans, citing a poignant legend about a vanished Native American tribe, whose only "survivor" was one old parrot; the bird still spoke their language, but no one could understand it because the entire tribe had been wiped out. Since parrots, like other birds, have almost no colon, their longevity seemed to support Metchnikoff's theory of aging. One seventy-year-old Amazonian parrot in his own lab was brightly feathered and as interested in females as ever, showing "no age-related changes except that his character had turned nasty." A female of the same species, dead at eighty-one, furnished cellular support for Metchnikoff's theories: its brain, unlike those of younger parrots, had been ravaged by phagocytes.

But more than birds, what intrigued Metchnikoff most were the bodies and souls of aging humans. He organized supplies of tissue samples of the old deceased from the morgue, scoured the newspapers for reports on unusual longevity, and sought out the elderly wherever he went to learn how they coped with life's approaching end.

Dealing so closely with life's finale compelled him to tackle philosophical questions. "Perhaps I'll suddenly take up philosophy in my old age," he wrote to Olga, who had gone to paint landscapes on the northern shore of France. He began to focus on core issues that had fascinated him since childhood. Where do we come from? Where are we headed? Why does life end in death? Now he was at last ready to tackle these questions head-on.

29

THE NATURE OF MAN

In contemplating matters of life and death, Metchnikoff discovered a tragic paradox that certainly applied to himself if not to everyone else: the closer one gets to life's end, the more one wants to live. As a result, he noted, people end up dying just when their "life instinct" is the greatest. "Our strong will to live runs counter to the infirmities of old age and the shortness of life. That's the greatest disharmony of human nature," he proclaimed in his first philosophical book, *The Nature of Man: Studies in Optimistic Philosophy*, published in Paris in the spring of 1903.

If people attained a "natural" span of life, uninterrupted by disease, malnutrition, or accidents—he estimated such a life span to be "significantly above a hundred years"—no one would fear death, he claimed. Not only that, people would die willingly. They would develop an "instinct" for death, just as they have a need for sleep. Metchnikoff assured the reader, "This instinct must be accompanied by marvelous sensations, better than any other we are capable of experiencing. Perhaps the anxious search for the purpose of human life is nothing but a vague yearning for this anticipation of natural death. It must be akin to the dreamlike sentiments of young virgins that precede true love."

Alas, he was the first to admit that evidence for the hypothetical death instinct was hard to come by, even though he kept searching, frequenting the crowded hospice for elderly indigent women at the Salpêtrière, the enormous Parisian hospital housed in an austere stone complex with a monumental domed chapel. It was in vain. "They all yearned for a long life," he wrote in *The Nature of Man* about the elderly he met, adding with dismay, "In the Salpêtrière, the main wish of the women aged ninety or more is to reach the age of one hundred; the desire to live is pervasive." To his disappointment, even the sick elderly did not want to die—they only wanted to get better.

Metchnikoff the atheist went as far as to search for "evidence" of his death instinct in the Old Testament: it describes Abraham and other patriarchs as dying "old and full of years." Metchnikoff's interpretation: "This phrase, which sounds so strange to our ears, probably refers to none other than the death instinct developed in well-preserved old men who had reached ages from 140 to 180 years."

Dying too soon, he argued, was only one of many disharmonies plaguing *la nature humaine.* Because natural selection, the main engine of evolution, has been too slow to adapt the human body to a changing environment, humans were encumbered by many organs that had outlived their usefulness—from the harmless hymen, foreskin, and tailbone to such harmful ones as the colon, the appendix, and wisdom teeth. Yet another major disharmony, to his mind, was maladapted sexual development, including the persistence of sexual desire into old age. "Old men, incapable of either inspiring or satisfying desire, often fall prey to their own amorousness and unassuaged passions. . . . Sexual disharmony can also occur between persons of different sexes. Its precocious emergence in males compared to females often leads to marital discord." He then quoted contemporary doctors who believed that masturbation at a young age could "ruin health, mental ability, and even life." Finally, he exclaimed, "It is remarkable how enormously the disharmonies of the reproductive functions in humans differ from the perfect adaptation of this same function in most plants!"

With his insistence on human disharmonies, Metchnikoff was being true to his own polemical self; he was challenging the man he called "the greatest living philosopher." Victorian intellectual Herbert Spencer thought that evolution spelled progress in and of itself, claiming that in almost every sphere, life was evolving toward greater complexity, which he equated with superiority. In *The Nature of Man,* Metchnikoff vehemently disagreed. "For Herbert Spencer, the great complexity of life in modern civilization is a sign of progress, but I believe that is not true," he wrote. In nature, he continued, complexity at times signifies disaster. Cockroaches and scorpions are better survivors than certain mammals. They had been around since the dawn of evolution, whereas a few species of apes had during this period become extinct, despite being nearly as complex as humans. "Consequently, there is no blind striving toward progress in nature," he stated.

On the other hand, as is obvious from *The Nature of Man*'s upbeat subtitle, he was convinced people could overcome even the worst disharmonies—with the help of science. "Human constitution is mutable and can be altered to benefit humanity," he confidently declared in the book. "The first imperative is to try to restore the correct course of human life; that is, to turn disharmony into harmony—orthobiosis," he wrote, coining his own term for "correct living," from the Greek *ortho*, "correct," and *bios*, "life."

With this rather vague notion of a "correct course of life," he had finally managed to define life's purpose, a definition for which he had fruitlessly, and desperately, searched in his youth. "The purpose of human existence lies in going through a normal cycle of life, leading to a loss of the life instinct and a painless old age, bringing about a reconciliation with death." In future writings, he was to clarify that "merely living longer" was not enough. The goal was to go through "a progressive evolution of the life instinct until the emergence of the death instinct."

These ideas about poorly adapted humans, and particularly the notion of the death instinct, bring to mind the writings of Sigmund Freud, about ten years Metchnikoff's junior, who at the turn of the twentieth century was inventing a new, conflicted view of the human psyche in Vienna. Freud even used some similar terms, among them his controversial *Todestrieb*, or "death drive," sometimes translated as the "death instinct." But Freud's works had not yet produced a major impact by the time *The Nature of Man* came out. And he was to define his *Todestrieb* only in *Beyond the Pleasure Principle* in 1920—that is, when Metchnikoff had been dead for six years. Besides, Freud saw *Todestrieb* as a product of a mind permanently at war with itself, not at all the epitome of serenity Metchnikoff envisaged.

Freud also mistrusted the notion of natural death, familiar to him from poetry and the work of certain biologists. (Metchnikoff is not among those he cited.) "Perhaps this belief in the incidence of death as the necessary consequence of an inner law of being is also only one of those illusions that we have fashioned for ourselves 'so as to endure the burden of existence,'" he wrote in *Beyond the Pleasure Principle.*

Still, in 1934 Scottish physician Al Cochrane, a younger contemporary of Freud who knew him in Vienna, published an essay, "Elie Metchnikoff and His Theory of an '*Instinct de la Mort,*'" in the *International Journal of Psychoanalysis,* pointing to "some striking similarities" between Freud's and Metchnikoff's "death instincts"—"in the idea of a 'latent' instinct controlling the length of human life . . . in the use of biological evidence to elucidate psychological problems . . . in the idea of psychical disharmony in general which could be solved by the application of the results of scientific study." Cochrane also drew other parallels, probably originating in the intellectual climate of the era, between Metchnikoff and Freud. "Both dismissed religion as worthless and philosophy as of secondary importance. Both emphasized, though to a varying extent, the importance of sex in all fields of life. Both denied the existence of a tendency toward progress in nature. Both agreed in believing in the sole importance of science as a means of obtaining knowledge for the purpose of intellectual understanding," Cochrane wrote.

Cochrane even noted in his paper, "I took the liberty of asking Professor Freud if he knew of any contacts direct or indirect that might have existed between him [and Metchnikoff]. He answered in the negative." When Cochrane's essay was reprinted in the *International Journal of Epidemiology* in 2003, a modern psychotherapist wrote in an accompanying commentary, "Freud wanted to be a scientist but was a storyteller. Metchnikoff was a distinguished scientist who, perhaps, wanted to tell stories that would comfort him in old age and in his own extinction."

Metchnikoff did much more than tell stories. In the final chapter of *The Nature of Man,* written when he was approaching sixty, he called for more research into old age, of the sort he was conducting in his own lab, coining two new terms still in use today—from the Greek word *geron*, "old man," and from Thanatos, a personification of death. "It is most likely that a scientific study of aging and death, giving rise to two new branches of science—gerontology and thanatology—will lead to significant changes in the course of the final stage of human life," he wrote. "The prospect of making old age easily bearable and of prolonging human life is not mere utopia."

In that chapter, he conjures up a vision of the future that does sound decidedly utopian. Aware of their lives' purpose, people are far happier than they were in the past. The able-bodied, energetic elderly take over politics and the running of society from the less experienced younger politicians. Finally, the need for everyone to complete a full life cycle boosts morality, requiring that "people help one another much more than they do now." The book ends with a bequest to future generations: "If it's true that one cannot live without faith, let this faith be in the overriding power of knowledge."

Promptly translated into several languages, *The Nature of Man* enjoyed tremendous popularity around the world, going through five editions in Metchnikoff's lifetime and remaining in print a hundred years later. When it first appeared, P. Chalmers Mitchell—the secretary of the Zoological Society of London and editor of the English translation—went as far as to declare it "one of the most important scientific productions since the *Origin of Species*."

Yet even in Great Britain, where Metchnikoff's book was perhaps most enthusiastically received, reviewers voiced criticisms. "We will not quarrel with him for the paucity of authentic facts which support his extraordinary theory—he has, we believe, discovered one melancholy old gentleman in whom advanced years have developed the instinctive love of Death," the *Times* of London conceded. "There is no greater biologist among us. But as a philosopher we may confess him too rash, too hasty, too impassioned, to enjoy the freedom of the meditative mind. . . . M. Metchnikoff strikes us sometimes as childishly aggressive in his materialism." Many reviewers, indeed, were offended by his insistence that only material existence, not spiritual, was real and that death brought with it "total annihilation."

As had happened throughout Metchnikoff's life, the book revealed that his strengths became his flaws when taken too far. His relentlessness had been a must when he defended phagocytosis, but it could push him to the verge of absurdity in defending his philosophical theories—for example, when he enlisted the biblical patriarchs as research subjects.

Metchnikoff himself was unfazed by the critiques of his philosophical views. Not only that, he occasionally sounded almost nostalgic for the much more virulent attacks he had sustained in the immunity war. For instance, he provocatively opened one article on aging by quoting German scientists who believed his theory was *auf der Luft gegriffen*, "taken out of thin air."

Such objections galvanized Metchnikoff into action, just as they had in the immunity war. He began planning a follow-up book to rebuke his critics, launched new experiments in the lab in support of his theories, and—ever a pragmatic utopist—focused on elaborating the practical guidelines of orthobiosis. It's not enough to define the purpose of life as a full cycle, he wrote in a preface to one of the book's later editions: "One must also show how to achieve this normal cycle, overcoming all the obstacles in its way."

30

PAPA BOILED

VISITORS WHO STOPPED BY METCHNIKOFF'S lab at the Pasteur Institute around lunchtime could find him in his study, at a massive oak desk by a tall window, hovering like a sorcerer—some said like Dr. Faust—over a Bunsen burner to toast his bread or sterilize his knife, fork, and spoon. He abided by strict hygiene to prevent the entry of harmful germs into his gut, eating nothing raw and drinking water that had been filtered, then boiled. He plunged strawberries, dates, and other fruit—even peeled bananas, which he thought were insufficiently protected by their thick skins from microbes—into boiling water for several minutes before eating them.

His motivation was not only to slow aging but to avoid infectious diseases. Among the latter he counted cancer, a prescient if exaggerated belief. (We now know that a few cancer types can indeed be spurred on by viruses or bacteria.) He also excluded from his diet substances he thought could be harmful. He abstained from alcohol and avoided coffee, opting for weak teas from different blends shipped from Russia, and called smoking "a disgusting and stinking habit." Food, he believed, should be simple and consumed in moderate amounts. And moderation, in turn, would benefit society as a whole by helping to distribute wealth more equally. Today his overall approach, the emphasis on preventing disease through healthy living, sounds exceptionally modern.

Wherever he went, Metchnikoff preached what he himself practiced. When riding a train or an omnibus, he would explain to housewives how to keep their food free of germs. Entering a grocery store, he pleaded with the owner to protect the cheeses from flies by using a metal mesh and to serve the groceries, especially those to be eaten uncooked, by tongs rather than by hand. Once, he was appalled upon entering one of the largest patisseries in Paris to buy candy for children. "An impeccably dressed saleswoman took the candies by hand to place them in the box," he wrote later in an unpublished article, "A Bacteriological Excursion Through Paris." "I noticed that she was pale and had a swollen gland beneath her elegant collar. She constantly coughed without turning her face away from the candy." When the saleswoman came to Metchnikoff's lab several days later at his insistence, so many tuberculosis bacilli were found in her sputum she had to be hospitalized. "Often, the simplest and easiest measures can avert the risk of infection," Metchnikoff summed up.

Resentment and ridicule were often his reward. Bakers were angered by his claims that the Saint-Honoré cake, a classic French dessert of puff pastry and cream, was dangerous because it was filled with uncooked beaten egg whites. His dire verdict on grapes—he said they were covered with so much dust that eating them was "suicidal," not to mention that their seeds could cause appendicitis—brought upon him the wrath of the French Society of Viticulturists, echoed by the daily *La Justice*: "If one can't eat fruit, drink unboiled water, or kiss those you love, it's not certain that life would be worth living." When Metchnikoff went food shopping, a task he never entrusted to the impractical Olga, women selling produce on the sidewalks near the Pasteur Institute shouted, "Monsieur *le professeur*, come buy my vegetables, I've poured boiled water over them this morning." But as soon as he turned away, they told regular clients, "I just say that to please him." Evidently aware of his nickname Père Metch, they called him behind his back Père Bouilli, or "Papa Boiled."

Back in the lab, Metchnikoff tried to learn as much as possible about the germs that, despite all preventive measures, did manage to penetrate the body. Pasteur had believed that in the human gut, microbes were of great service or perhaps even indispensable for digestion. Metchnikoff launched pioneering studies into gut microbes that became a central theme in his lab: determining what effect these microbes really exerted on the body. He believed that under certain circumstances, particularly during disease, some of them could cause a detrimental process akin to rotting that he called "intestinal putrefaction." The issue was controversial; some scientists believed no such process took place in the gut and if it did, that it wasn't necessarily harmful. Others argued that even consumption of rotten food caused no harm, citing the peculiar preference for rotten fish and meat known to exist among people in Greenland, Polynesia, and a few other parts of the world.

A young doctor, Henry Tissier, had just then joined Metchnikoff's lab after completing a thesis at the University of Paris on the intestinal flora of sick and healthy newborns. Metchnikoff assigned him to study the rotting of meat, which, he thought, could shed light on the supposed intestinal putrefaction. As Tissier and another researcher report in the *Annales*, in one of the experiments they swallowed pieces of rotten horse meat to study its impact on their gut chemistry. (The reader may be relieved to know they did not get sick.) These were the kind of heroic acts Metchnikoff inspired in his assistants.

Next, Metchnikoff instructed Tissier to study another type of chemical transformation of food: fermentation that causes milk to turn sour. Half a century earlier, it had been Pasteur who discovered that fermentation was the work of living organisms. Tissier wanted to find out why the microbes, instead of causing the milk to rot, rendered it sour but free from unpleasant types of decomposition. Was milk more stable than meat because of the sugar or certain proteins it contained?

Metchnikoff's team supplied evidence for a different explanation: sour milk owes its stability to the lactic acid that microbes produce from milk sugar in the course of the fermentation. The acid acts as a preservative, keeping the milk palatable while preventing undesirable decomposition. It kills the microbes that could cause decomposition. (In fact, if the acid is produced in high amounts, it can end up killing those same microbes that secreted it.) Rot-causing germs failed to grow when placed in the same broth with milk-souring organisms, but when the scientists added soda, which neutralizes acids, the rot-causing germs began to multiply immediately. "The milk microbes consistently produce large amounts of acid, which hinders the activity of the putrefaction bacteria," Metchnikoff wrote in *The Nature of Man*, which he was just then preparing for publication.

He was well familiar with the use of acids to prevent food from rotting. Russians consumed pickled cabbage and cucumbers by the barrel, especially to make up for the lack of fresh produce during the long winters. He also knew that in some countries, meat was preserved by storage in sour milk or whey.

Preserving the milk itself by souring had in fact been practiced since antiquity in warm regions the world over. The end product depended on the cultures used and was sometimes alcoholic, if the cultures contained yeast that fermented part of the milk sugar into alcohol.

The liquid, slightly alcoholic koumiss was made from mares' milk in the steppe of southeastern Russia; the jellylike *leben*—from the milk of buffaloes, cows, or goats in Egypt; the creamy *yahourt* or *yoghourt*, spelled differently depending on the region—in Turkey and the Balkans. In the Caucasus, hospitable shepherds served travelers a type of acid-fermented milk called *kefir*, prepared with "grains"—yellowish spongelike clumps of bacteria and yeast—the shepherds believed had been a gift from Allah. To make the *kefir*, they filled leather tubes sewn from calf stomachs with milk and a few "grains" and placed them in the sun by the side of the road, where children and passersby kicked them occasionally to provide the necessary stirring. "If a man cannot reconcile himself to sour milk, he is not fit for the Caucasus," a British mountaineer warned in an 1896 book about the region.

As for western Europeans and Americans, short of those who embarked on exotic travels, they generally tasted sour milk only when they got sick. Koumiss and *kefir*, for instance, were thought to be good for digestion and helpful against anemia, tuberculosis, and other wasting diseases. In 1875 a London manufacturer advertised his koumiss in the *Times* of London as a beverage that could "nourish persons when nothing else will, excelling cod liver oil as a restorative of strength and weight."

Back in Russia, in his twenties, Metchnikoff had indeed bought koumiss from Tatar merchants for his first wife, Ludmila, to try to cure her tuberculosis. Even though the koumiss had wrought no miracle, in Paris he started drinking sour milk regularly in his early fifties, around 1898, when he had begun to worry about his health, imagining that his kidney problems signaled a terminal illness. By the time his assistants launched the sour milk experiments, he had been drinking it daily for nearly five years, believing it stimulated digestion and the cleansing of the body.

Tissier's experiments on the cleansing effects of lactic acid bolstered his belief. "These findings explain why lactic acid often stops certain instances of diarrhea," he wrote in *The Nature of Man*. If sour milk generally blocks rotting, he reasoned, perhaps it could block toxins in the gut. "Clearly, the slow intoxication, which weakens the resistance of our noble elements, can be arrested by sour milk, preferably of the type that doesn't contain alcohol," he wrote.

Just then one Professor Leon Massol, a former Pastorian who had moved to the University of Geneva, provided Metchnikoff with a sampling of Bulgarian sour milk known in Bulgaria as *yahourth* or *kisselo mleko*. Stamen Grigoroff, a Bulgarian student in Massol's lab, had found it to contain rodlike organisms, later called *Bacillus bulgaricus*, that generated particularly large amounts of lactic acid—twenty-five grams per liter of milk. No less significant was Grigoroff's observation that a surprisingly large number of centenarians lived in the region of Bulgaria in which the *yahourth* constituted the main staple of diet.

Life expectancy in much of western Europe and North America had soared in the preceding century, from just over thirty years in 1800 to around fifty in the early 1900s. But the increase had occurred because there were fewer deaths in childhood and young adulthood, whereas for older people, on average, mortality rates had dropped only slightly, if at all. The maximum life span hadn't changed either. People who passed the hundred-year mark remained sufficiently rare to make headlines. In fact, numbers of centenarians began to decrease in the twentieth century, not because there were fewer of them, but because the numbers were being revised downward due to the improving accuracy of statistical data. Stories about large numbers of centenarians were more likely to circulate about mountainous villages or other regions with no reliable birth registers.

But Metchnikoff's desire to find a cure for aging was so strong, he did not seem to care whether the stories about Bulgarian centenarians were reliable or not. Suddenly, it all came together for him, as it had all those years earlier in Messina. On the one hand, the large intestine was infested with "bad" germs. On the other hand, milk-souring microbes blocked the growth of such germs. And the sprightly, elderly Bulgarians consumed extraordinary amounts of sour milk. Metchnikoff's hypothesis: milk-souring germs could help counter the harmful gut microbes responsible for aging.

And as he had done in Messina, he was about to infuse modern meaning into yet another health concept formulated by Hippocrates: achieving health through diet. There was no proof, only an idea. Neither Metchnikoff nor anyone else could have imagined the revolutionary impact it was to have on the diets of future generations worldwide.

31

THE BUTTER-MILK CRAZE

IT'S RARE TO BE ABLE to trace a global dietary trend to a single event, but the modern yogurt industry was arguably born in the lecture hall of the Society of French Agriculturalists in Paris on June 8, 1904—the day Metchnikoff delivered there a public lecture, "*La Vieillesse*" (Old Age).

He spoke with his usual ardor, which seems to have only intensified as he had aged. "To the casual stranger who had just dropped into the room, he might have been some fiery anarchist orator denouncing the sins of society in violent language, such was the force of his gesture and the vivacity of his whole manner," recalled a journalist who attended one of the lectures Metchnikoff delivered in his sixties. "To give additional emphasis to what he was saying, he thumped the table, nodded his head continually, waved his arms, and swayed himself to and fro with astonishing animation, whilst every now and then he would pull a huge handkerchief out of his pocket and mop the perspiration from his face." Once, after Metchnikoff gave a talk at a medical congress, a member of the audience commented with a broad smile on his delivery, "The man is possessed of a demon."

In the "Old Age" lecture, Metchnikoff described the sorry lot of the elderly everywhere—from the archipelago of Tierra del Fuego, where in times of famine, according to lore, old women were killed and eaten before dogs were, because "dogs capture seals, whereas old women don't," to western Europe, where the elderly committed suicide and fell victim to murder much more often, he claimed, than the rest of the population. He then shared with the audience his hypothesis of aging as "a chronic infectious disease"—that it resulted from the toxins released by harmful microbes in the large intestine. To demonstrate his point, he had brought with him a pair of animals differing sharply in the way they had aged: a decrepit seventeen-year-old dog—an unfortunate owner of a large intestine—and a feisty septuagenarian parrot, which was lacking this organ. "The dog seemed offended by the comparison, looking as if it wanted to devour the parrot," *Le Figaro* commented later.

Metchnikoff urged his listeners to avoid eating raw food as much as possible. "It's a source of all sorts of wild microbes," he said. "Even if raw fruit and vegetables, such as strawberries, cherries, lettuce, and radish are washed, they retain some of the dust, earth, and sometimes excrement." But avoiding harmful germs was not enough; beneficial ones had to be cultivated in the intestine, such as the one isolated from the Bulgarian sour milk. "Interestingly, this microbe is found in the sour milk consumed in large amounts by the Bulgarians in a region well-known for the longevity of its inhabitants. There is therefore reason to suppose that introducing Bulgarian sour milk into the diet can reduce the harmful effect of the intestinal flora," Metchnikoff said.

The following day, the lecture was front-page news and the talk of Paris. Metchnikoff had been careful to present his ideas as a hypothesis, never affirming a direct link between sour milk and longevity, but by now he should have known better—all his caveats were edited out of the euphoric press reports. Published in Paris as a separate brochure and soon translated into English, for months the lecture continued to make headlines in France and all over the world, amid reports on the naval battles of the Russo-Japanese War, the opening of the third modern Olympics in St. Louis, and the victory of Theodore Roosevelt in the US presidential election.

In its article "Eliminating Old Age," *Le Temps* wrote, "Those of you, pretty ladies and brilliant gentlemen, who don't want to age or die, here's the precious recipe: eat *yaghourt*!" In England, the *Pall Mall Magazine* ran a long interview with Metchnikoff on sour milk and his theory of aging under the headline CAN OLD AGE BE CURED? In the United States, the *Chicago Daily Tribune* printed an article headlined

SOUR MILK IS ELIXIR, with the subheading SECRET OF LONG LIFE DISCOVERED BY PROF. METCHNIKOFF. It explained that "any one desiring to attain a ripe old age is recommended by Prof. Metchnikoff to follow the examples of the Bulgarians who are noted for their longevity, and who consume large quantities of this cheap and easily obtained beverage."

Before long, the mailman was unloading at the Pasteur Institute piles of letters asking for information about the new elixir of youth. Unable to answer them all, Metchnikoff, who had been planning to devote a chapter of his upcoming book to milk-souring microbes, decided to publish this chapter ahead of time as a small brochure. *Quelques Remarques sur le Lait Aigri* (*A Few Remarks on Soured Milk*) first saw light in France in the fall of 1905 and later appeared in several English translations in London and New York.

In addition to explaining his theory of aging, the brochure contained instructions for preparing healthful sour milk. Since raw milk could contain dangerous germs, yet lengthy boiling ruined its taste, Metchnikoff proposed a compromise: boiling the milk, preferably skimmed, for just a few minutes. After cooling, it had to be seeded with a pure dose of beneficial microbes, covered, and left for a few hours in warm temperature. The resultant sour milk could be eaten at any time of the day, with meals or separately, Metchnikoff advised. "Its use at five hundred to seven hundred milliliters per day regulates the bowel movements and improves the secretion of urine," he wrote.

Trying to counter the sensational press reports, Metchnikoff concluded the brochure with a disclaimer: "Clearly, we do not look upon the milk microbes as an elixir of longevity or a remedy against aging. This question will be resolved only in a more or less distant future."

It was too late. The cautionary statement couldn't quench the soaring thirst for sour milk. Besides, Metchnikoff had actually fanned the mania by mentioning in the brochure that he himself, having regularly consumed sour milk for seven years, was "satisfied with the results," as were many of his friends who had followed his dietary example. Parisian newspapers revealed that among these was none other than Monsieur Roux, famous throughout Europe for his work on the antidiphtheria serum. He was preparing his own sour milk by placing a tightly closed jar with germ-seeded milk under the blanket in his bed overnight.

That the cream of France's scientific elite embraced skimmed sour milk added to its cachet. After all, in 1904, the year of the "Old Age" lecture, France's Academy of Sciences made Metchnikoff a corresponding member, and that same year the Pasteur Institute appointed him deputy director—assisting Roux, who had become the institute's director.

Besides, sour milk had a compelling advantage over any of the other life-extension methods attempted over the centuries. Involving no risky injections or costly concoctions, it was available to all.

Doctors from the world over were telegraphing the Pasteur Institute with requests for cultures of milk-souring germs or even personally traveling to Paris for a consultation. Among the latter was a fiftyish American with a bushy mustache who ran a holistic sanitarium in Battle Creek, Michigan, in which he advocated his own version of healthy living based on a vegetarian diet, exercise, and sexual abstinence. (He used pure carbolic acid to discourage female masturbation.) Impressed by Metchnikoff's ideas and by the pitcher of sour milk he saw on his desk, Dr. John Harvey Kellogg later wrote in his book *Autointoxication* that Metchnikoff had "placed the whole world under obligation to him in his discovery that the flora of the human intestine needs changing." Returning home with cultures from the Pasteur Institute, Kellogg made sure each of his patients received a pint of sour milk. Half was eaten, and the other half was administered by enema, "thus planting the protective germs where they are most needed and may render most effective service."

Doctors everywhere started prescribing sour milk—also referred to as "butter-milk," "Oriental curdled milk," "yoghourt," "yoghourth," "yaghourt," or "yohourth"—to treat diarrhea in babies and various digestion problems in adults, but they also gave it to patients to help prevent gout, rheumatism, and the clogging of arteries. In Carlsbad and other popular European resorts, sour milk became the treatment in vogue. A medical review article in Great Britain titled "On the Use of Soured Milk in the Treatment of Some Forms of Chronic Ill-Health" recommended giving patients sour milk in preparation for surgery, as a disinfectant of the digestive tract. One Glasgow physician treated women with chronic gonorrhea by rinsing their vaginas with whey. A doctor in Berlin was delighted to discover that typhus bacteria disappeared from the stools of his patients after he had given them milk curdled with Metchnikoff's cultures. "The milk was very sour, but this didn't prevent the patients from taking it effortlessly for several months," he wrote to Metchnikoff.

In its commentary on "the butter-milk craze," the *Lancet* claimed that "since Metchnikoff began to publish his investigations and the butter-milk theory became popular, much clinical evidence has been forthcoming." And as with every remedy, doctors warned about side effects. "It may be well to direct the attention of those who wish to try this sour-milk treatment to the fact that they should assure themselves beforehand that they are fit subjects for it, and they should therefore consult a medical man," the *Lancet* cautioned. The *British Medical Journal* chimed in, "Yoghourt can be used for an indefinite time without harmful results if the dose be not too large, 1 kg a day should not customarily be exceeded."

Ignoring the warnings, the public craved the milky fountain of youth. In March 1905, a shop near Théâtre du Vaudeville in one of the most fashionable spots in Paris began to offer "Maya Bulgare," *maya* being a term used to describe certain ethnic yogurt cultures. Ads in *Le Figaro* invited all Parisians who cared about their health "to taste the delicious Bulgarian curdled milk that the illustrious Professor Metchnikoff has recommended for suppressing the disastrous effects of old age." Top-hatted gentlemen and frothy-gowned ladies strolling the nearby *grand boulevard* dropped by the shop to savor the exotic food on the spot, or to buy a few portions to take home for what the ads called their "five o'clock yoghourt."

"In mundane circles," reported the Paris correspondent of the *Boston Medical and Surgical Journal*, later renamed the *New England Journal of Medicine*, Metchnikoff's theories "have had a *succès fou*, and as they fitted in exactly with their wishes, which were to remain young and beautiful on the female side, and vigorous on the male, everybody in this town has since then been taking Metchnikoff's milk with a fervor proportionate to the scientific authority of its promoter."

Within months of Metchnikoff's lecture, milk-souring germs had blossomed into an international business. Pharmacies throughout Europe and the United States were offering Bulgarian cultures in the form of tablets, powders, and bouillons—to be consumed as is or used to make sour milk at home in regular jars or in special incubators, marketed under such brand names as Sauerin, Lactobator, or Lactogenerator. And because the Bulgarian bacilli were said to produce lactic acid not only from milk sugar but from ordinary sugars, food manufacturers responded with chocolate, candy, and other sweets carrying the germs. An ad in the *Times* of London featured Massolettes, milk-souring bacilli "packed in a dainty chocolate cream," apparently named after the Geneva professor Leon Massol, in whose lab the bacilli had been first isolated.

Metchnikoff constantly had to intervene to prevent his name from being abused. In February 1905 poster-sized ads appeared on boulevard de Clichy, close to Montmartre, to promote milk "ferments" that "delay aging and eliminate disease, according to Doctor Metchnikoff of the Pasteur Institute," with Metchnikoff's name spelled in large block letters. Metchnikoff had to appeal to a court for the ads to be removed, as he would have to do on numerous occasions over the years to stop vendors from claiming they had trained under him or had received his endorsement.

Metchnikoff himself had never dreamed of making money from research. Not only that, he had worked at the Pasteur Institute for years without as much as requesting a salary. But soon he would have a secret reason to associate himself with a commercial undertaking.

32

A TRUE MALADY

In 1903, after fifteen years in Paris, Metchnikoff had finally realized his fantasy of living in a peaceful little community. He and Olga had moved permanently to Sèvres, about six miles west of Paris, where five years earlier they had bought a house that at first served them as a summer dacha.

Today Sèvres, a gentrified and serene old town, is a chic Parisian suburb. Its name evokes the famous pricey porcelain manufactured by a local factory since the eighteenth century, often decorated by a characteristic light-blue color, *le bleu de Sèvres.* Stone mansions along the westbound road remind the visitor that the town had once been a popular stopover for French aristocrats on their way from Paris to the royal residence in Versailles. The street names are a record of famous former residents: Balzac, Gambetta, Corot, Eiffel, Metchnikoff.

When the Metchnikoffs settled here, it was already an easy commute from Paris, reachable by two different train lines. A steam locomotive leaving from the Montparnasse railway station, conveniently near the Pasteur Institute, took Metchnikoff to the lower, commercial area of Sèvres, from which he could climb up to the upper, more affluent part of town in which he lived. Another train, from the Saint-Lazare station, brought him directly to upper Sèvres, where on his way home he passed the birches, maples, and aspens of the formerly royal park of Saint-Cloud, once a favorite painting spot of Camille Corot and other French landscapists.

The Metchnikoffs' narrow, three-story house on 28 rue du Guet, built of reddish-brown grit stone with white window frames, a high porch accessed by two side staircases, and vines crawling up the walls, stands behind an iron-grid fence on close to half an acre of land, whose northern side abuts the park. The plot, now divided between two families, is enclosed by a wall of bushes, bamboo, and ivy, as well as trees, among them yews thought to date back to Metchnikoff's time. Top-floor windows facing the street open onto a superb panorama of Sèvres and the surrounding hills, with a cone-shaped turret of a stone watchtower rising nearby amid greenery and tiled roofs.

The Metchnikoffs' residence, in the words of a friend of theirs, was "a small oasis of science and art." In its grassy backyard smelling of mint, they built a vine-covered studio—with a large window open to the northern light, so that it be free from inconvenient shadows—in which Olga painted and sculpted. Metchnikoff got up at five o'clock in the morning and wrote or read at home before taking a train to Paris. Upon returning home around seven in the evening, he first stopped by the kitchen, whose provision was his responsibility (the vegetarian Olga, by some accounts, hardly ate anything at all), then received guests and read to Olga out loud or listened to her play his favorite Mozart and Beethoven on the piano. ("If I weren't a scientist I would have been an orchestra conductor," he once said.) "How pure the air is, how green the grass! What peace! You see, if I hadn't spent the day in Paris, I wouldn't have appreciated the charm and quiet of Sèvres!" he used to tell Olga.

When *The Nature of Man* came out in 1903, he dedicated it to Olga. At around the time of its publication, his feeling of guilt toward her, to which he had once admitted in his correspondence with her, must have reached an intense pitch. "I dedicate this book to my dear, beloved girl, in the hope of attracting her to optimism," he wrote in ink on the copy he gave her, a hint that Olga wasn't too cheerful at the time. Soon after the book appeared, he fathered a child outside his marriage—or so at least he believed.

The child, Lili, was officially his goddaughter, born to Pasteur Institute scientific illustrator and photographer Émile Rémy and his wife, Marie. From the start, Metchnikoff doted on Lili visibly more than he did on all other children of his colleagues and friends to whom he had served as godfather. Olga acknowledged in his biography, "He particularly loved one of his goddaughters, the little Lili; he became so attached to her because of her exceptional kindness, and the affection she exhibited toward him virtually from the cradle." Metchnikoff's fondness for Lili was "common knowledge" at the Pasteur Institute. After visiting the institute, the American journalist Edwin Slosson wrote in his book *Major Prophets of Today*—profiling Metchnikoff and five other scientists, philosophers, and writers—that the Metchnikoffs "have no children, but he has a godchild to whom he is devoted."

More than 150 of Metchnikoff's letters preserved by Lili, as well as her unpublished memoirs in the Pasteur Institute archive, make it possible to trace Metchnikoff's extraordinary relationship with her parents and herself. It had begun in November 1894 when Marie, then still single, had come to seek Metchnikoff's advice about her anemia. She was twenty-two, an intelligent and excitable young woman with neatly parted hair, delicate features, and large expressive eyes, who spent all her free time reading. Just arrived in Paris from her native village in France's Massif Central, she must have been awestruck by the world-famous scientist and grateful for his continued advice concerning her condition.

Metchnikoff, then forty-nine, had recently returned from the Budapest Congress, ready for a truce in the immunity war. There are no signs he was ever a womanizer in his youth, but with age he seems to have grown more susceptible to female charms, or perhaps more emotionally available as his scientific battles subsided. "When I examine a myxomycete [a slime mold], it's as if I'm looking at a beautiful woman," he once said. His interest in sex seems to have also grown with age, as suggested by his plans in the last few years of his life to write a book, *Studies of the Sexual Function.*

It was probably in Metchnikoff's lab that Marie met the tall, dark-haired Émile Rémy, two years her junior. After they were married in November 1901, Metchnikoff took charge of them, advising them on financial and medical matters and exchanging frequent letters with them whenever he or they went away, getting progressively more involved with their lives after Lili was born in July 1903. He urged them to give her healthy food, to sprinkle antitetanus serum on all her cuts, and generally "to take care of our dear little Lili, our beloved, adorable angel, as the apple of your eye."

When the Rémys settled in Sèvres, for a while renting a house on the same street with Metchnikoff and Olga, on 14 rue du Guet, Metchnikoff stopped by every morning before taking a train to Paris, and sometimes in the evening too, on his way home. Holding Lili on his knees, as she slipped her arm around his neck, he could spend hours reading to her out loud. Whenever he had visitors from abroad, Marie brought Lili over to be introduced to his guests. Once, upon receiving her photo in the mail, he showed it to everyone at the Pasteur Institute, just like a proud parent. When Olga went away, he took his suppers at the Rémys', on certain occasions actually moving in with them for some time.

His book *The Prolongation of Life*, a follow-up to *The Nature of Man*, published four years after Lili was born, contains a veiled autobiographical passage, in which, as we have already seen, he describes—in the third person, speaking of "a close friend"—his own transformation from a painfully sensitive, suicidal young man into a much happier, if mollified, mature adult. Perhaps "the friend's" most dramatic change was in his attitude toward children. The young man thought it was criminal to bring them into the world, whereas the mature adult takes in them his greatest delight. "The simplest phenomena, such as the smile or the babbling of a baby, the first words or statements pronounced by a child, became a source of true happiness for him," Metchnikoff wrote, clearly referring to Lili.

"In other words, from a pessimist, my friend turned in his later years into a convinced optimist, though this didn't prevent him from suffering greatly, mainly due to illness or grief of those close to him," he wrote in conclusion.

His happiness was indeed marred by constant concern about Lili. When she got sick, Metchnikoff was inconsolable. Years later, in her unpublished memoirs, she recalled how he had reacted when as a little girl she was in bed with a sore throat. "He kissed my hands and cried. I caressed his head, I was so upset to see him like that." When describing the last few years of his life, Olga wrote, "His love for children, even for his favorite goddaughter, no longer gave him joy, he worried so much about them."

It's not that Metchnikoff's concern was entirely unfounded. Suffice it to recall that at the time, there was hardly a family that hadn't lost a child to disease. Metchnikoff wrote that in France of the early twentieth century, one in four children died in his or her first year of life. Still, his angst seems exaggerated even by the standards of his time. Whenever the Rémys went away, he anxiously awaited their letters, losing sleep as he imagined that something bad had happened to Lili. "Please write every day, I feel sick and unable to work when I don't hear from you," he implored them. Once, describing his constant worry about Lili's health in a letter to Marie, he admitted, "My love for Lili is a true malady."

Metchnikoff's desire to secure Lili's future even pushed him, shortly after his "Old Age" lecture, to sponsor a commercial enterprise, Société Le Ferment. Friends warned him against taking such a risk—in France, a scientist was not supposed to get rich from research—but he couldn't forgo this unique opportunity to provide for Lili, then just over a year old. So after consulting the Pasteur Institute legal adviser, Metchnikoff agreed that the company could use his name. He granted this right free of charge but made one condition: Émile Rémy became the company's founding partner.

Operating off boulevard du Montparnasse, not far from the Pasteur Institute, Le Ferment, proudly advertised everywhere as "the sole provider of Professor Metchnikoff," sold sour milk and pills called Lactobacilline. Their package insert stated that Lactobacilline offered "a method for replacing harmful intestinal flora with one that is beneficial"—that is, "composed of different rigorously selected lactic ferments. These ferments have been studied by Professor METCHNIKOFF, who has established their perfect safety."

The endorsement produced the desired effect. Le Ferment was so successful it even expanded overseas. In the United Kingdom, it granted a concession to the London Pure Milk Association, which sold a type of yogurt named Lactobacilline Milk. An American subsidiary, the Ferment Company of New York, marketed the following: Lactobacilline pills; Bacillac, a yogurt-powder supplement; and Lactofermentine, a suspension of the Bulgarian bacilli in broth, for use against abscesses and wound infections.

Several years later, with the rise of the xenophobic sentiment in France, this commercial success was to expose Metchnikoff to public condemnation. When friends reminded him he should have been more careful about lending his name to a commercial enterprise, he said he had no regrets. "He answered that the choice had been between the welfare of an entire poor family on the one hand, and gossip on the other," Olga wrote in his biography. "He never hesitated to provide help, even at the risk of poor consequences for himself." Olga's romantically heroic description of her husband revealed nothing about the suffering she herself endured because of Metchnikoff's preoccupation with Lili.

33

ABSURD PREJUDICE

ONCE, WHEN LILI WAS PUNISHED for calling a schoolteacher *Madame* instead of *Mademoiselle*, Metchnikoff stormed into the school, infuriated, grabbing Lili by the hand to lead her away. He often intervened with her upbringing with the authority of a parent. His relationships with his other godchildren, including Lili's younger brother Elie, named after him, never came close to this level of involvement. He was irate when Lili's piano teacher said she should practice three hours a day. "I protest against this exaggeration," he wrote. "I live and breathe anew," he wrote to Marie after receiving a letter that Lili was fine. "I think about you every day and see you in my dreams almost every night," he wrote to Lili herself.

A century after Metchnikoff had written these words, his yellowed letters convey an emotional intensity that does seem to expose a parental sentiment. But was Lili, who looked very much like her mother, indeed Metchnikoff's child? Having no children with his wife, it's not inconceivable that at fifty-eight, he might have simply developed an inordinate fascination with a goddaughter. After all, he made no attempt whatsoever to hide his adoration of Lili from Émile. Many of his doting letters about her are openly addressed to both Marie and Émile, *mes bien chers amis*. When addressed to Marie alone, they invariably end with "best regards" or "a friendly handshake" to "dear Émile." As for the profuse terms of endearment he applies to Lili, they could very well have been his style, familiar from his published letters to Olga.

One letter, though, ends all doubts. For once, Metchnikoff clearly intended it for Marie's eyes only. Using the intimate *tu* instead of the distant *vous* of other letters, he wrote to arrange for them to meet in private every day, perhaps even twice a day. Marie would walk Lili to school in the morning, Metchnikoff would follow by tram. They would get together "in the lower part of Sèvres," evidently in a hideout familiar to both. Instead of a handshake to Émile, the letter ends with the words, "Tender kisses to my dear Marie, whom I love with all my heart."

So even though there is no way of knowing for sure whether Lili was Metchnikoff's biological child, one thing is certain: he had good reason to believe she was.

It is unknown when Metchnikoff's love affair with Marie began. His first surviving letters to her date from the beginning of her marriage in 1901. At around the same time, Metchnikoff's own marriage, by then more than a quarter of a century old, was going through a rocky stage. Ever since the immunity war had wound down, Olga felt he no longer needed her, despite his assurances to the contrary. With increasing frequency she started going away to paint in the country, spending a great deal of time in a forest retreat community known as the Colony, west of Paris. Metchnikoff tried to reassure his "beloved fugitive." "I love you very, very deeply, you are making a big mistake in failing to acknowledge this," he wrote to Olga two months before Lili was born. "How groundless it is to think about separation," he wrote on another occasion.

"And what about your attachment to Roux?" he exclaimed in yet another letter. "What have *I* got of this sort?" In fact, it would have hardly been unexpected if Olga had turned to Roux for consolation. In photographs taken when Lili was a small child, Olga, around fifty, appears to be anything but a neglected wife, looking more sensuous than ever with a stylish bouffant and an arched back, her youthful figure clad in a tight-fitting dress.

But Metchnikoff's correspondence with her and others discloses evidence of her suffering. Whether or not she knew Metchnikoff was having an affair, she was mainly jealous of his attentions to Lili. She could compete with another woman, but not with a little girl. It must have been a nightmare to watch her husband suddenly revel in fatherhood, real or imagined, when she herself was past childbearing years.

"It's hard for me, knowing that I'm returning home without being sure you'll be pleased; I feel so acutely that all your interests now lie with other families rather than with me," she wrote to Metchnikoff when Lili was three. "Of course I'm attached to the child, I enjoy listening to her chatter and following her development," he protested. "But what's this got to do with my being interested in her more than in you?" He tried to hide from Olga his "terrible anxiety" about Lili's well-being, perhaps recalling guiltily that when he and Olga had married, he was still adamantly against having children as part of his pessimist philosophy.

From Metchnikoff's correspondence with the Rémys, it's clear that Émile didn't know about the affair, or at least acted as if he didn't. The only surviving hint at an imbalance in his marriage is an isolated sentence in unpublished notes from an interview with Lili by a Pasteur Institute historian: "Émile loved Marie more than she loved him." As for his relationship with Metchnikoff, it's not inconceivable that Émile's judgment was muddled by his gratitude for all that his famous patron had done for his family.

What does seem surprising, at least at first glance, is that Metchnikoff, a man of otherwise impeccable integrity, keenly interested in ethics and morality, should have found himself at the head of not one but two romantic triangles—one with Olga and Roux, the other with Marie and Émile. How could he have ended up not only betraying a wife who had devoted the most productive decades of her life to his career but also cuckolding an unsuspecting young man who took pride in being his protégé?

Perhaps the answer may be sought in the nihilist idealism of his youth, coupled with his own utopian philosophy. Metchnikoff viewed contemporary social conventions as imperfect and transient. He repeatedly insisted that in all spheres of life, including personal relations, science would ultimately usher in an improved societal order. Unlike the most militant Russian nihilists, he never called for marriage to be abolished, but like them, he thought the future would bring about more felicitous ways for men and women to bond, though he never specified exactly how. He may have believed, or at least wanted to believe, he wasn't acting immorally, only ahead of his time.

He criticized the biographers of Robert Koch, who avoided mentioning Koch's affair with an actress less than half his age, which led to his divorce from his first wife, "as if there was something about his family life that could denigrate the memory of the great scientist." In his book *Founders of Modern Medicine: Pasteur, Lister, Koch*, written two years before his death, Metchnikoff defiantly tells the story of Koch's marriage to the actress, discarding as prejudice all criticism of Koch's affair. In the future, he declared, "the absurd prejudice, which today governs even the relatively independent minds, will be eliminated." Metchnikoff wrote that other professors condemned Koch for his romance because they "couldn't forgive him his scientific superiority." He concludes with a statement he might very well have applied to himself: "With time, when marital relations are organized incomparably better and simpler than they are today, Koch will be judged far more fairly by posterity than he was by his contemporaries."

With that, Metchnikoff's views on women were far from forward-looking. In fact, they were outrageously sexist. He believed genius was a male trait, once telling a journalist that women would never amount to much as scientists, even if exceptions did exist, "just as there are bearded ladies." Being personally acquainted with two outstanding women scientists—the mathematician Sofia Kovalevskaya, wife of a friend of his, and Marie Curie, who worked in Paris at the same time he did—failed to change his mind. He cited them among the "exceptions." On another occasion, he insisted that the lack of prominent women had nothing to do with discrimination against them, since, to his mind, they had no major accomplishments even in areas traditionally open to them, such as music or cooking. "When I need a good dinner I'll invite a male cook," he told a reporter. He evidently wasn't bothered by the fact that in progressive circles of his time, such opinions had already long become passé, even if in a large part of society they were still very much the norm. His one saving grace is that he believed in equal educational opportunities for women, extending his help to female students in Russia and France on many occasions.

As for his affair with Marie, he might have derived yet another excuse for his own behavior from his belief, stated in his later writings, that amorous exploits had a stimulating effect on work. "It would be very important for humanity's well-being to possess in-depth information about the love lives of great people," he wrote in his memoirs about one Russian physiologist. Love and sexual desire are intimately linked to creativity, he argued in the chapter called "Goethe and Faust" in *The Prolongation of Life*, citing examples from the lives of great people, all of whom, unsurprisingly, were older men like himself: Victor Hugo, Henrik Ibsen, Richard Wagner. In the same vein, he provided in that chapter an original explanation to the second part of Goethe's *Faust*, considered one of the most enigmatic works of world literature. This second part, according to Metchnikoff, was nothing but a story of Goethe himself, in his seventies, falling passionately in love with a pretty nineteen-year-old called Ulrike, an "elderly love" that Goethe, fearing ridicule, had supposedly disguised so much as to make the work incomprehensible.

It's possible that further in that same chapter, Metchnikoff makes an unspoken reference to his own affair. "Not only artistic creativity but other forms of genius are closely connected with sexual activity," he wrote. "Scientists are no exception to this rule." Goethe's Dr. Faust is a case in point. "A scholarly doctor, having acquired all human knowledge, remains dissatisfied, finding solace in the beauty and loveliness of a young girl, Margaret, with whom he falls passionately in love," Metchnikoff wrote about Dr. Faust. He then wonders about ways in which this affair affected Dr. Faust's work, and ways in which the work affected the affair: "It would be most interesting to determine the psychological dynamics of this transition from scientific laboratory studies to a setting involving Margaret."

If Metchnikoff's plans to write *Studies of the Sexual Function* were inspired by his relationship with Marie, then, as he had done with aging, he extended his own preoccupations to thoughts about all of humanity. We know from the preface to the book, the only part he managed to complete, that he hoped scientific research would uncouple sex from its traditional linkage to shame and sin. Religion had made this link, he argued, only because "since ancient times, sexual function had been closely associated in the human mind with venereal diseases"—an association that science, he was sure, was about to sever. He wrote, "It's unacceptable for the modern, civilized world to be guided by principles dating from the time when total ignorance prevailed in medicine and hygiene."

34

RETURN OF A PSYCHOSIS

METCHNIKOFF DID NOT GET A chance to write the book about liberating sex from shame, but for a short while he did manage to help make sex safer. His original goal had nothing to do with safety—or with sex, for that matter. He began to experiment on syphilis because it resembled aging in one of its cardinal symptoms: a hardening of the arteries. "A man is as old as his arteries," was a saying he recalled. On the other hand, he believed that aging, like syphilis, was a microbial disease. So he set out to understand how syphilis hardens the arteries.

Syphilis was viewed as such a disgrace its name could not be mentioned in polite company. Instead, the French used the nickname "*avarie,*" "damage." Historically, people in different countries had blamed it on the neighbors, calling it the "German disease" in Spain, the "French evil" in Germany, the "Polish disease" in Russia, and the "Chinese pox" in Japan. The damage it caused was horrific. It began with genital sores but was often followed by painful, foul-smelling abscesses on the lips, eyes, and all over the body, and in later stages could sometimes cause paralysis and death.

It was a prevalent scourge too. One of Metchnikoff's brothers, the lawyer Nikolai who had died a few years earlier at age fifty-six, had been a sufferer. And syphilis exacted its toll on public health not only as a sexually transmitted disease. In an 1895 pamphlet published in St. Petersburg, medical authorities expressed the concern that the Russian people might degenerate because of a *sifilizatsiya* of the population in rural areas, where one case sometimes infected an entire village. Peasants spread the infection through open sores, by eating and drinking from shared dishes, kissing one another during festivities, and feeding their own and other people's children with chewed-up bread. The only treatment was mercury, a poison that wasn't much better than the disease itself. "A night with Venus and a lifetime with mercury," the saying went. Medicine was making no progress. Not only was the germ that causes syphilis unknown, but the disease was difficult to study because common laboratory animals didn't contract it.

Metchnikoff decided to try and infect with syphilis the closest relatives of humans: monkeys and great apes. The first obstacle was money. To procure a chimp or orangutan from French or British colonies cost an exorbitant 1,000 to 2,000 francs, roughly half the annual salary of a Pasteur Institute research assistant. The animals themselves were mistreated during shipment and often died quickly because they were unaccustomed to the climate. In a move applauded by the French press, Metchnikoff and Roux donated their own prize money to the cause—Metchnikoff's 5,000 francs, from the Moscow Prize (so named because of the source of the funding) he was awarded in 1903 at the Fourteenth International Medical Congress in Madrid, and Roux's whopping 100,000 francs, from the Osiris Prize. Another large donation, 30,000 francs, came from a Moscow philanthropist.

Success came with the very first try. In July 1903 Metchnikoff and Roux arrived at the Academy of Medicine in Paris accompanied by a porter who carried a large case. It contained Edwige, a young female chimpanzee suffering from "a disease unknown in the forests of the Congo, where she had been captured," according to a compassionate report in *Le Matin* titled THE "DAMAGE" IN APES—A SENSATIONAL ANNOUNCEMENT AT THE ACADEMY OF MEDICINE. Metchnikoff proudly showed the academy members the ugly ulcer in Edwige's pubic region, identical to the chancres formed on the private parts of syphilitic humans. He had managed to infect her with syphilis. (At the time, the antivivisection movement, particularly in England, was already voicing criticism of animal experiments; Metchnikoff frequently defended such experiments in his writings as "the only means for science to make serious progress.")

In the coming three years, he was to infect more than fifty chimps with syphilis, whose disease proved to resemble that of humans most closely, as well as an entire primate zoo—orangutans, gorillas, large Guinea baboons with hairless rumps, and nearly eighty macaques of various species, among them several pink-faced rhesus monkeys. He was to publish five scientific papers in the *Annales* on the course of syphilis and its various aspects, as well as two reports in the bulletin of France's Academy of Medicine.

The primates created an unexpected interference. The unending stream of visitors to Metchnikoff's lab now included journalists and other Parisians who wanted to see "the monkeys." On top of that, self-imposed duties as the most famous Russian living in Paris already encumbered Metchnikoff.

When Olga and friends organized the benevolent Montparnasse society in 1903, he added to its prestige by becoming its president. When a Russian sculptor was unsure her works would be accepted to an exhibition, he enlisted the assistance of his acquaintance Auguste Rodin. When a young revolutionary had a problem with her passport, Metchnikoff intervened on her behalf with the Russian consulate. And when the Russian Higher School of Social Sciences opened across the street from the Sorbonne to provide the hundreds of young Russians in Paris with an opportunity for a college-like education, he was asked to serve as its president, in addition to teaching a course on biology.

Among the school's visiting faculty was a short, balding man in his early thirties, with a reddish beard and vaguely Mongolian features, called Vladimir Ilyich Ulyanov. Roaming Europe to try to foment a Russian revolution from abroad, he taught a course on Marxist agrarian management under the pseudonym Ilyin, but at about the same time he also started calling himself Lenin. The older brother whom he had idolized, Alexander Ulyanov, a Narodnaya Volya member hung for an attempt on the tsar's life, had been a biologist. So it must have been with particular zeal that Lenin, in vain, kept trying to recruit the great biologist Metchnikoff to the revolutionary cause. "Every time I meet our friend Ilya Metchnikoff, I thank him for his yogurt and reproach him for staying away from the social issues of mankind," Lenin once told a fellow Russian expatriate, Chaim Weizmann, an organic chemist and future first president of Israel, who was briefly training in microbiology at the Pasteur Institute.

The number of Russians turning to Metchnikoff soared in the wake of the failed revolution of 1905. When in January of that year, on what was to be known as the Bloody Sunday, dozens of thousands of factory workers, holding religious icons and singing "God Save the Tsar," tried to approach the Winter Palace in St. Petersburg to submit a petition to the tsar, the imperial guard shot at the unarmed crowd, killing or wounding hundreds. Workers' strikes and mass protests spread throughout the country and were brutally suppressed. By the time the revolution ended in the summer of 1907, thousands of insurgents had been executed or exiled. Some fled Russia. Many of them ended up in France, soon bringing the total number of Russians there to about thirty-five thousand, most of them living in Paris and on the Côte d'Azur.

In the wake of these events, waves of exiled Russians besieged Metchnikoff's lab, asking for money, counsel, or help with finding work. "They are dressed very poorly and seem in a sorry state, but what can I do for all of them, especially since they come right off the street, just because 'everyone knows me,'" Metchnikoff lamented in a letter to Olga. On another day, he reported to her, "Several Russian ladies came today for advice: one suffers from mouth sores, another from hair loss, a third one has shiny skin on her face, and so on. Now that such masses of Russians have fled abroad from the revolutions, I have a feeling many of them come to me because they have nothing else to do, just to gawk at a famous scientist."

It was not only the Russians who came by the dozen. The lab was supposed to be open to outside visitors in the afternoons, but Metchnikoff never turned people away at any hour, curious to hear the news they brought from all over the world. "The discoverer of the phagocytes is known today in every laboratory; and when a young chemist comes from England or Germany to Paris, his first curiosity is to make the acquaintance of the Russian savant whose Quixotic campaign against old age has fired the general imagination," the *Times* of London wrote. One of Metchnikoff's younger associates recalled, "His laboratory at the Pasteur Institute was a salon of a new genre, a scientific European caravanserai. Scientists, statesmen, artists, journalists (so many), actresses, singers, society ladies, cranks, the sick who came to ask for Rejuvenating Water or simply a vaccine, the tormented who came to confess, and so many Russians: the tsar's ministers, politicians, travelers, refugees, fine gentlemen, balalaika players, thin and eager students, and so many 'cadgers' of all nationalities."

Among the Pastorians, too, Metchnikoff had a knack for drawing a crowd. Whenever a group of people gathered somewhere at the institute, it was a safe bet he was at its center. While in his lab, Metchnikoff, who read everything and was legendary for his encyclopedic memory, served as a reference desk for colleagues and students. It was not without a certain coquetry that he told them where exactly to look for the information they needed, including the part of the page on which the desired article began. When he himself needed to remember a name or a number, he noted it on a scrap of paper he then immediately discarded, the writing duly recorded by his brain.

In letters to Olga, Metchnikoff repeatedly complained about "an orgy of interferences." "An indescribable number of Americans showed up today, all speaking their inscrutable English," he wrote in one letter. Then there was "the mad doctor" who kept coming to share his "discovery" of cancer parasites. "He's completely pig-headed but so sincere, I feel sorry for him," Metchnikoff moaned to Olga. To work for a few hours in peace, he locked himself up in the institute's library. He looked forward to Sundays and holidays as opportunities for working without interruptions. His hospitality, a wonderful trait, turned against him when overly extended.

It was because of all the disruptions, he was sure, that he had missed discovering the syphilis microbe. His lab was among many in the world that were trying to find it. The microbe was elusive because it was difficult to make it visible with the standard dyes. Finally, in 1905 the international race was won by two German scientists. The microbe turned out to be shaped like a very thin coil, which looked pale even when scientists did manage to color it with a special dye.

At the outset, its discoverers had a hard time convincing the scientific community; in the preceding decades, a syphilis germ had been "found" at least two dozen times. Then Metchnikoff's primates came to the rescue. At the request of the German scientists, he confirmed the presence of the coiled microbe in infected chimps and macaques, providing crucial support for its role in syphilis.

A note of bitterness at having himself failed to find the germ can be detected in one of his letters to Olga. Referring to constant disruptions, he wrote, "Working under such conditions, no wonder I hadn't discovered the syphilis spirilla."

But he did make an important discovery about syphilis. He showed that the disease did not spread immediately throughout the body, which suggested that if caught early, it could be blocked. In monkeys and apes, such blocking proved possible by rubbing the site of infection with an ointment containing calomel, a mercury compound that was then widely used in France to treat anything from toothache to tuberculosis. But could humans be protected in the same manner?

Immediately, a number of young men declared their willingness to expose their genitals to syphilis in the name of science. The same "psychosis" (as Metchnikoff himself had called it) that had once possessed him during his cholera studies descended upon him anew. Forgetting that he had vowed never again to experiment on human beings, he turned from a paternal figure into a fanatic patriarch.

On February 1, 1906, in the presence of three medical doctors, Metchnikoff dipped a scalpel into discharge from a syphilitic ulcer, then used this same scalpel to make six scratches on the foreskin of one Paul Maisonneuve, a medical student preparing a thesis on the prevention of syphilis. An hour later, calomel ointment was rubbed into the scratches for five minutes. One can only imagine Metchnikoff's agony when he examined the purulent vesicles that formed on Maisonneuve's foreskin in the following few weeks. Should the young man come down with syphilis, he risked being disfigured by ulcers or even dying. Three months later, the doctors present at the experiment examined the student. All his scratches had healed, they pronounced. He had not contracted syphilis.

Newspapers published a photo of Maisonneuve, "the hero who has risked his health for the progress of science." Metchnikoff, on the other hand, came under fire. Several physicians criticized him for experimenting on another human being rather than on himself. His rebuttal was that he had absolute faith in his method because it had worked in chimps and that throughout his career he had already conducted on himself numerous risky experiments, which indeed he had.

Other critics accused him of encouraging promiscuity. It was "immoral to let people believe they could go to Cythera [the island of the goddess of love] with impunity" and "indecent to give them a means of wallowing in debauchery," pronounced the *Tribune Médicale*. Undaunted, Metchnikoff argued at every opportunity that morality was on his side. "Is it immoral to spread a means that rescues families from being debased, children from being born with hereditary syphilis, and an entire people from going into a decline?" he exclaimed in a lengthy interview with *Le Matin*.

"Everything having to do with sexual function is considered so base and shameful that people prefer to avoid discussing it, at the risk of causing great harm by this hushing up," he wrote in the preface to *Studies of the Sexual Function*. He further called for sex education in schools, a radical idea for his time. "He came to entirely revolutionary conclusions regarding education and marriage," Olga wrote in his biography. She doesn't specify what kind of revolutionary ideas he had about marriage, but he considered his work on the prevention of syphilis as a step toward sex being no longer perceived as sinful.

For all the controversy, *la pommade de Metchnikoff*, or "Metchnikoff's ointment," was soon saving soldiers and sailors from contracting syphilis. "Prevention of sexually transmitted diseases in the army seems to have entered a new phase," *La Presse Médicale* wrote in 1907, reprinting a pamphlet in which France's deputy minister of war instructed army physicians to educate soldiers about "applying calomel ointment to parts of skin or mucous lining exposed to contamination" to prevent syphilis and ordering infirmaries to dispense five-gram doses of the ointment in small wooden boxes, as needed.

But within a few years, calomel ointment as a safe-sex measure was to be pushed aside. In 1910 Paul Ehrlich was to release the arsenic-based Salvarsan, a drug working on his "magic bullet" principle that specifically targeted the syphilis germ. It brought amazing relief at least in some syphilis sufferers. Quickly becoming the most widely prescribed drug in the world, Salvarsan remained the preferred treatment for syphilis until penicillin became available in the 1940s. As in the immunity war, Ehrlich had upstaged Metchnikoff yet again.

35

BIOLOGICAL ROMANCES

At about the time Metchnikoff was losing the race for the syphilis germ, the relevance of his old immunity research soared suddenly, helping to pave the way for his Nobel Prize. Sir Almroth Wright, a provocative and influential British physician who admired both Ehrlich and Metchnikoff, devised a new therapy that relied not only on Ehrlich's ubiquitous antibodies but also on Metchnikoff's neglected phagocytes.

Wright, a bespectacled man with bushy eyebrows and a mustache, was even more combative and prickly than Metchnikoff. Once, when asked by the chairman of a medical committee whether he had anything to add to his testimony, he replied, "No Sir, I have given you the facts—I can't give you the brains." Nicknamed by detractors "Sir Almost Right" or "Sir Almost Wrong," he argued that physicians would one day become "immunisators" because bacterial infections would be treated primarily with vaccines, not drugs. This view was soon to be shattered by the antibiotics revolution—ushered in by, among others, his own pupil, penicillin discoverer Alexander Fleming.

One of Wright's greatest achievements was his successful campaign to have British troops preventively vaccinated against typhoid fever. Another was the discovery of antibodies that enhance phagocytosis. After noticing that phagocytes swallow the germs faster in blood serum than they do in a neutral solution, he revealed the existence of antibodies that attach themselves to bacteria or other foreign cells, thereby making the invaders more susceptible to being devoured by phagocytes.

He called these antibodies *opsonins,* from the Greek *opsonein,* "to render palatable." "The blood fluids modify bacteria in a manner which renders them a ready prey to phagocytes," he wrote in a 1903 paper. Continuing the tradition of British researchers attempting to reconcile the two rival immunity camps, his finding provided a concrete link between humoral and cellular immunity.

The therapy Wright devised based on his discovery—injections made from cultures of offending bacteria accompanied by the monitoring of the "opsonic index" (that is, the measurements of opsonins in the patient's blood)—for a while gained substantial popularity among physicians worldwide. The monitoring was intended to reveal whether the patient's phagocytes were succeeding at "beating the germs," explained a Johns Hopkins University physician writing in the *New York Tribune.* In the full-page, optimistically titled article END OF DISEASE, devoted to Wright's method, he went on to say, "The opsonic index affords the physician a certain report on his patient's condition, and he is thus enabled to work intelligently instead of gropingly, as formerly."

When in May 1906 Metchnikoff was invited by the Royal Institute of Public Health to deliver the prestigious Harben lectures in London, he used the opportunity to spend an entire day with Wright, visiting his lab and seeing the cancer and leprosy patients he was treating with the opsonin method in the department he headed at St. Mary's Hospital. Years later, at a celebration of Wright's seventy-fifth birthday, Wright insisted that authentic praise should always be qualified by reservations, quoting with particular relish Metchnikoff's words about himself: "Wright is the sort of man who has very good original thoughts—but he has also many thoughts that are only original."

Metchnikoff didn't class the opsonin theory with the "only original" ideas, but he argued that even if the preparatory action of opsonins was absolutely necessary, it was still "far less important than that of the phagocytes." But his scoffing of opsonins, a British discovery, did not in the least detract from the veneration British scientists and physicians continued to lavish upon him. Throughout his London stay, they greeted him with utmost splendor everywhere, which never swayed him either from his modest ways ("The Englishmen were exceptionally gracious, even too much so," he wrote to Olga after Royal Institute members had gathered in a hall to welcome him with applause before his first lecture) or from his sartorial negligence ("I was the only one in a frock coat, but this no longer embarrasses me in the least," he reported after a dinner at which all the other men wore tails).

Yet another sign of appreciation came in the form of a flattering offer of a professorship from the University of Cambridge, "wither your renown has long since preceded you, and where you will be sure of a most cordial welcome." For a short while, Metchnikoff considered accepting. In the wake of the 1905 upheavals in Russia, he was worried about losing Olga's country estate, their main source of income. "We'd be in poor shape, with our dealings in science and painting, if we weren't secure about the future," he once wrote to Olga. At the Pasteur Institute, he was earning very low wages in his new capacity as deputy director—4,000 francs a year, about a quarter of the income from the Russian estate. (He had earned no wages at all in his first sixteen years at the institute.) But after receiving an assurance from a senior Russian politician that landowners would never see their estates expropriated in Russia, Metchnikoff decided to stay at the Pasteur Institute, the scientific home to which he was so attached.

In the fall of that same year, Metchnikoff was awarded the Copley Medal, the highest decoration of the Royal Society of London, "on the ground of the importance of his work in zoology and in pathology." And as if all these honors were not enough, artistic Britain recognized him with an "honor" of its own. George Bernard Shaw spoofed his phagocytes in his play *The Doctor's Dilemma,* a satire of the medical profession that in November 1906 opened in London's Royal Court Theatre.

In the preface to the play, Shaw called the British health care system "a murderous absurdity." It could only benefit, he suggested, from being ridiculed in the literary tradition started by Molière. He then disparaged fashions in medicine, mockingly referring to "the dramatic possibilities" held by the opsonins: "Sir Almroth Wright, following up one of Metchnikoff's most suggestive biological romances, discovered that the white corpuscles or phagocytes which attack and devour disease germs for us do their work only when we butter the disease germs appetizingly for them with a natural sauce which Sir Almroth named opsonin." Shaw ended the preface with a list of pithy conclusions, one of them probably an allusion to Metchnikoff's longevity theories: "Do not try to live forever. You will not succeed."

As for phagocytes, in *The Doctor's Dilemma* they fill the slot of a medical mantra, part jargon and part nonsense; just as in Molière's medical farces, doctors use fake Latin to cover up their ignorance. The only one to question their worth is the surgeon Cutler Walpole, a man of forty who "seems never at a loss, never in doubt: one feels that if he made a mistake he would make it thoroughly and firmly." Walpole tells a patient, "If the phagocytes fail, come to me. I'll put you right."

What Metchnikoff thought of the play is unknown, but he surely agreed with Shaw that opsonins were being overrated. After returning from England, he wrote in a scientific review on immunity, "Despite the activity of opsonins, the bacteria stay alive, multiplying and maintaining their virulence. In contrast, phagocytes put an end to the vital energy of the bacteria they consume."

Today we know that opsonins indeed facilitate phagocytosis, even if their role is nowhere near as central in immunity as Wright believed. But more than a hundred years ago, researchers lacked the tools for following up on Wright's pioneering glimpse into the interactions among different mechanisms within the immune system. His opsonin therapy soon fell out of favor as too complicated. Still, for a while it revived the interest in phagocytes, so much so that the Nobel Committee was to write in its 1908 evaluation of Metchnikoff's immunity research, "These achievements of his have lately (through the research on opsonins etc.) been given a new vitality and importance."

36

MAIS C'EST METCHNIKOFF!

"TABLES RAISED FROM ALL FOUR legs, movement of objects from a distance, hands that pinch or caress you, luminous apparitions"—that was how Pierre Curie described a séance of a popular spiritual medium in a 1905 letter to a friend, commenting that "the phenomena we saw appeared inexplicable as trickery." Unlike Curie, Metchnikoff, after attending séances by the same medium and others in Paris, concluded that they were nothing but trickery. "They all ended in failure," he reported happily.

The new popularity of the séances was but one sign of a widespread disillusionment in science. Metchnikoff believed it had set in because science had been unable "to curtail human suffering"—that is, it hadn't lived up to its promise of freeing humanity of all major diseases. Nearly three decades after the discovery of the tuberculosis germ, for example, no cure for this scourge was yet in sight. Even the triumphs of science seemed fraught with perils; they had spawned technologies that brought with them overcrowding and pollution. At the dawn of the twentieth century, as confidence in science waned, fascination with spiritualism and the occult had soared in French society.

If Pierre Curie and a few other leading scientists of the day were willing to give paranormal powers the benefit of the doubt, it was because many of the recent scientific advances seemed so bizarre—the radioactive elements isolated by Curie himself and his wife, Marie, the X-rays discovered by Wilhelm Roentgen, the weird quantum world exposed by Max Planck. These perfectly scientific findings could help legitimize a belief in mysterious, invisible forces. If X-rays were real, why not telepathy?

Metchnikoff, in contrast, was appalled by society's slide into mysticism. He was particularly troubled by a prevalent view, diametrically opposed to his own, that science didn't run sufficiently deep to explain the world. It was with obvious despair that he described a lecture he attended by the charismatic French philosopher Henri Bergson, who argued that instinct and intuition were far more crucial for understanding reality than science and rationalism. What vexed Metchnikoff most was the cult following Bergson attracted.

The lecture, at the Collège de France, was so overcrowded with "people of all ages, from youth of both sexes to gray-haired old men and women," wearing "all types of dress, including immense mushroom-like hats," that listeners struggling to get in broke off the glass in the entrance door. After the lecture, Metchnikoff, always on the lookout for new insights into death, approached Bergson, a tall, slender man with a high forehead and a compact mustache, asking him to clarify his oft-quoted poetic passage about humanity galloping "in an overwhelming charge" to overcome "formidable obstacles, perhaps even death." Bergson replied that he hadn't yet given much thought to ideas behind this passage but was planning to do so in the future. "Death is a very interesting experiment," he explained.

In 1907, the same year that Bergson's most famous work, *Creative Evolution,* came out, Metchnikoff published his second philosophical book, *Études optimistes,* which in the English translation was given a longer, attention-grabbing title: *The Prolongation of Life: Optimistic Studies.* Bergson's book proffered a fashionable philosophy. Metchnikoff's epitomized the other extreme. Once again, he was going against the grain, defending an unpopular position at the time—that is, the supremacy of science.

The Prolongation of Life was to turn out to be impressively forward-looking in defining numerous medical and social concepts, such as the relative roles of heredity and the environment in longevity and the societal challenges of caring for the elderly. "Obviously, the prolongation of life should go hand in hand with preserving strength and the capacity for work," Metchnikoff proclaimed. When one of the book's twenty-first-century editions came out, a 2005 review in the *Journal of the American Geriatrics Society* called it "a surprisingly relevant book today."

Throughout the book, Metchnikoff addresses questions still hotly debated in modern science. What determines the length of human life? What's the maximum age a person can reach? He conceded that many historic supercentenarian stories were myths, such as that of the 185-year-old Kentigern, the patron saint of Glasgow, but treated more recent accounts less critically, particularly the famous legend about Old Tom Parr, the Shropshire peasant who supposedly died in London in 1635 at the age of 152 years and nine months. This piece of "evidence" sufficed Metchnikoff to reach his oft-quoted conclusion, "We can suppose humans can reach the age of 150 years."

The animal world also gave him much reason for optimism. There are creatures that don't age at all, he noted, referring to certain single-celled organisms that go through seven hundred divisions with no signs of aging in the older generations. He was particularly thrilled when a male turtle he had received as a gift sired baby turtles at age eighty-seven. "We can conclude that decrepitude, that is, premature aging, one of the major scourges of humanity, is not as deeply rooted in the organism of higher animals as might appear at first glance," he commented.

In fact, even in proclaiming himself an optimist, Metchnikoff, yet again, was going against the prevailing tide. At the time his book came out, optimism and pessimism were regarded as philosophical positions rather than attributes of temperament. In the words of a modern Oxford University scholar, "The contest between optimism and pessimism was a prominent feature of the intellectual life," so defining where one stood on the issue was only natural. Pessimism was a much more trendy contemporary philosophy, particularly in France, so much so that a group of intellectuals in Paris created a Society of Optimists to preach cheerfulness. "Added to gloominess transmitted to us from ancient times we have, in a country like France, a system of pedagogy which places pessimism on a pedestal," one of the society's founders explained to the *New York Times.* "All our literature is permeated with this spirit.

Even Voltaire said, 'I do not know what eternal life is, but I do know that this one is a nasty practical joke.'"

Metchnikoff was once invited to preside over a banquet of the short-lived Society of Optimists, even if strictly speaking his worldview, with its emphasis on disharmonies, occupied a position more to the middle of the pessimism-optimism spectrum. A more extreme optimist would argue that the world was basically a good place and that in the course of history it was getting progressively better. Metchnikoff believed it was humanity's task to improve an imperfect world. He ends the last chapter of *The Prolongation of Life,* titled "Science and Morality," with the lofty admonition, "People are capable of great deeds. That is why it's desirable to change human nature, turning its disharmonies into harmonies. Only human will can achieve this ideal."

The Prolongation of Life received a great deal of publicity all over the world, as had Metchnikoff's previous philosophical book. In one of the most glorious reviews, *Le Figaro* announced rapturously that "by daringly planting himself in front of Old Age and Death in order to extract their secret, Metchnikoff resembles a distant son of [Oedipus], who once upon a time, at the gates of Thebes, had managed to solve the riddle and strike down the Sphinx." There were critical reviews too, but the general reaction was one of enormous interest, adding to Metchnikoff's already substantial fame.

"*Mais c'est Metchnikoff!*" (Oh, that's Metchnikoff!) exclaimed visitors to the 1908 salon of the Société Nationale des Beaux-Arts in the Grand Palais when they stopped before the immense allegorical painting *Toward Nature, for Humanity,* by the society's president Alfred Roll. In the painting, which today hangs in an amphitheater in the Sorbonne building, three of France's most famous scientists in academic mantles, Metchnikoff among them, march in a rocky valley, thrusting their arms toward a bright source of light, apparently representing knowledge.

Metchnikoff was featured in the canvas not just as a great scientist but as a great scientist counted among what the press called *personnalités parisiennes.* Along with ministers and diplomats, he was invited to presidential receptions for visiting royalty in the Hall of Festivities of the Elysée Palace. When *Les Flambeaux,* a popular new drama set in a research institute, premiered at the Porte-Saint-Martin Theater, it was Metchnikoff's glassware-filled Pasteur Institute laboratory that the producers sought to re-create on stage. A French naval expedition gave the name Metchnikoff Point to an extremity of the Pasteur Peninsula in one of the islands of the Palmer Archipelago off the coast of Antarctica.

For all his celebrity, Metchnikoff never gave himself airs. Leo Tolstoy, whose works of fiction he so admired, once said that each person is like a fraction: the numerator is what he is, and the denominator is what he thinks about himself; the greater the denominator, the smaller the fraction. By this measure, Metchnikoff would have made a large fraction. Despite his fame, he remained as nihilistically egalitarian as ever. Nor did his fame ever dampen the childlike curiosity that instructed all his research—or his equally childlike bluntness. One summer, when Manuel II, the young ex-king of Portugal, visited the Pasteur Institute, a group of scientists welcomed him in the garden. Metchnikoff, standing near the ex-king, looked at him with great interest, then, turning around, grabbed a colleague by the arm and declared, pointing a finger to Manuel's neck, "Tubercular ganglions!" "*Cela jetta un froid*" (That cast a chill), an institute researcher noted about this episode in his memoirs, using a French expression for a conversation stopper.

In a Paris boiling with artistic energy, the frequency with which painters and sculptors asked Metchnikoff to pose was perhaps the best measure of his star status; it was the closest equivalent to being stalked by paparazzi today. One of the most acclaimed of contemporary artists to ask Metchnikoff to sit for a portrait was the symbolist Eugène Carrière, who is now remembered mainly for his friendship with Auguste Rodin but who was at the time one of France's most famous painters. Metchnikoff granted such requests with self-humor, reporting to Olga when Carrière started making sketches, "I did my best to assume the pose he was after, with a book in one hand and a finger of the other hand in my mouth for biting a nail, but nothing worked."

In Carrière's portrait, Metchnikoff's head and upper torso, in shades of beige, appear shrouded in the artist's characteristic mist. Edgar Degas reportedly offered venomous explanations for the mistiness in Carrière's various paintings: "The model had moved," or "Someone had again been smoking in the children's room!" But Metchnikoff, habitually interpreting everything in terms of biology, thought Carrière painted monochrome, blurred canvases because his vision must have been impaired by kidney trouble.

Metchnikoff himself paid no heed to artistic Paris, which was all the more striking considering that the artistic scene burgeoned right around the corner from the Pasteur Institute, in the semirural Montparnasse area, which was on its way to replacing Montmartre as the artistic mecca. Making a detour from the institute to boulevard du Montparnasse, Metchnikoff walked by such headquarters of the café society as La Rotonde, in which Modigliani then used to sketch portraits of customers on the terrace, or La Closerie des Lilas, where Picasso hung out with poet friends.

In one of Metchnikoff's favorite spots in Paris, the botanical garden Jardin des Plantes, he might have run into yet another doyen of modern art, Henri Rousseau, making sketches for his *Combat of a Tiger and a Buffalo.* Having never left France, Rousseau painted tropical scenes inspired by the only jungle he had ever seen, the one growing in the Jardin. Within a short walk from the Pasteur Institute, Metchnikoff might have stopped to chat in Russian with a young Marc Chagall, Chaim Soutine, or other artists newly arrived from Russia to La Ruche, "the hive," an artist colony in the Montparnasse area next to rue de la Convention, where Metchnikoff did his grocery shopping.

Metchnikoff's own Russianness not only did not impede but perhaps even eased the French scientific establishment's embracing him. The French continued to be infatuated with all things Russian, despite a surging xenophobia. Metchnikoff, with his expressiveness and utopianism, fit the lyrical stereotype of the "Slavic soul"—one supposedly swayed by passions as boundless as the vastness of Russia—that came into vogue in France. It was precisely in such melodramatic terms that years later, a former student was to tell Metchnikoff's life story at a symposium on syphilis in Paris: "The Slav loves to despair: hasn't love pushed him to the brink of an abyss, hasn't he tried to end his life after losing a loved one, his first wife?"

Even though Metchnikoff remained a subject of the Russian Empire, the French claimed him as a glory of their country's science. In addition to earning a few of France's most prestigious awards and being elected to its Academy of Medicine, he was ultimately promoted from a corresponding member of France's Academy of Sciences to an *associé étranger*, the highest status he could achieve as a foreigner.

Nor did his sensational theories undermine his standing in the international scientific community, though it often came close to that. "Some of the more conservative temperament call [Metchnikoff] 'yellow,' even while admitting that his discoveries relating to the function of the white corpuscles in the blood as scavengers and as wardens against disease are unassailable," the *New York Times* wrote. He continued to be granted honorary memberships in scientific academies and societies in Russia, Europe, and North and South America, ultimately accumulating nearly a hundred, as well as diverse awards from all over the world, among them a Bene Merenti medal from King Carol I of Romania and an Albert Medal from Great Britain's Royal Society of Arts.

37

TRIUMPHANT TOUR

In early November 1908, Metchnikoff received a letter from Stockholm that was not entirely unexpected. He had been on an informal "waiting list" for the Nobel Prize from the time the prizes had been first awarded in 1901. Not only had he been nominated each year, but the number of nominations had been growing. This growth was spurred on by the flourishing of immunity research, which, in turn, led to an unprecedented assault on infectious diseases.

Following the success of serum therapy for diphtheria, the search was on for conquering all the major killers of humanity: tuberculosis, smallpox, the plague. Vaccines to prevent cholera, typhus, and tetanus were coming into widespread use. An immunological blood test for syphilis was developed. The discovery of opsonins generated the hope for new immunity-based cures. Finally, the Nobel Committee declared in its confidential evaluation of Metchnikoff's candidacy, "He has the great merit of having pioneered the modern research on immunology and to have for a long time led its development."

Metchnikoff was named laureate of the 1908 Nobel Prize in Physiology or Medicine. There was only one catch: he was honored jointly with Paul Ehrlich, "in recognition of their work on immunity."

When the prizes were presented in the Great Hall of the Royal Swedish Academy of Music in Stockholm on December 10, the anniversary of Alfred Nobel's death, Ernest Rutherford received the prize in chemistry "for his investigations into the disintegration of the elements," and Gabriel Lippmann received the one in physics "for his method of reproducing colors photographically." In a speech introducing the two laureates in medicine, Count Mörner, rector of the Karolinska Institute, declared, "Elie Metchnikoff was the first to take up consciously and purposefully, by means of experiments, the study of the question so fundamental to the question of immunity; by what means does the organism vanquish the disease-bearing microbes." But it was Russia's ambassador to Sweden, Baron Fyodor Andreyevich Budberg, who stepped forward on the red carpet to accept from King Gustaf V of Sweden, in Metchnikoff's name, a solid gold medal with Nobel's profile and an artistically designed diploma in a leather cover with an *EM* monogram encircled by a wreath. Metchnikoff himself did not attend the ceremony.

It is true that the Nobel Prize did not yet have its gold-standard cachet, but because of the caliber of the recipients, not to mention the sum of the award, it had already gained enormous prestige, making front-page news around the world each time laureates were announced. Metchnikoff's unusual decision to skip the presentation of this all-important award because of teaching duties—he was busy teaching a course at the Pasteur Institute, he explained in a letter to the chairman of the Nobel Committee—does sound like an excuse, and it probably was. A more likely reason was the vexing pairing with Ehrlich. Painfully aware that his phagocyte theory was being overshadowed by the rapid progress in antibody research, Metchnikoff evidently preferred to spare himself the humiliation of playing second fiddle to Ehrlich in Stockholm. Incidentally, Ehrlich, for his part, had his own reason to be mightily annoyed by their pairing: he believed he deserved a Nobel Prize all on his own.

Now that secrecy has been lifted from the Nobel archives of the early twentieth century, it emerges that in terms of the number of nominations, they both were strong candidates. From 1901 to 1908, Ehrlich had received sixty nominations; Metchnikoff had accumulated sixty-six, mostly from France, Russia, and parts of Poland that were then Austria-Hungary, but also from other countries, including the United States and even Germany. In the opinion of science historian Alfred Tauber, the Nobel Committee's deliberations suggest that "Metchnikoff was acknowledged as 'the founder' of immune research and Ehrlich as the better experimentalist."

From today's perspective, the committee showed remarkable foresight in recommending that they share the prize, even though caution had, in part, dictated the decision. If divided between the two of them, the committee wrote, the prize will not be "seen as guaranteeing the correctness of their respective theories, while at the same time recognizing the great service that they have undoubtedly both rendered to the development of immunology."

Surely, Metchnikoff was flattered by the Nobel Prize. Even if sour milk and calomel ointment had made him more famous than his immunity studies ever did, he knew what his true gift to humanity had been. But the prize came too late to provide him with unadulterated joy, a sentiment that was foreign to him in any event. By then, he had been engrossed in research on aging for at least seven years. Being hailed for his work on immunity was like learning he had won last year's race. He admitted in a letter to a former student in Russia that he had been much more honored by the Moscow Prize he had received in 1903, adding that "financially, the Nobel Prize is of course much more gratifying."

He and Ehrlich received 97,000 francs each, about $500,000 in today's money. At least for a while, he could be free of financial concerns. "I no longer need to write articles for popular magazines to make yet another one hundred francs, but can devote all my productivity to science," he wrote in the same letter to Russia. It is unknown what exactly he did with the prize money, but it was soon gone. Indirect evidence suggests he spent at least some of it on research, including the purchase of apes (his share of the Nobel Prize was worth, in theory, nearly a hundred chimps, but since he was no longer working on syphilis he didn't need as many). He gave some of the money away in charity and invested in zinc and gold mining in Mexico and elsewhere, encouraging Émile Rémy to follow his example. "The enterprise bore no fruit until the end of his life, and he was greatly distressed by having misled those who followed his advice," Olga wrote.

It was only in mid-May 1909 that Metchnikoff, accompanied by Olga, traveled to deliver the obligatory Nobel lecture in a cold and windy Stockholm. "They say here that the Swedish climate is divided into two: nine months of white winter and three months of green winter," he wrote to Marie and Émile. "The latter has yet to arrive, as the trees are still completely black and the bushes are barely beginning to sprout tiny buds." He devoted his Nobel lecture to the subject for which he had won the prize, but it would have been out of character for him to give a mere scientific overview.

Speaking in French, a decision he later regretted, echoing his habitual feeling of being misunderstood (*the Swedes know German better than French,* he wrote to the Rémys), he opened his Nobel lecture with an acknowledgment of having been honored "together with my excellent friend, Professor Ehrlich." (He may have been unaware that in *his* Nobel lecture, Ehrlich hadn't mentioned him at all.) Metchnikoff went on to polemicize as if his long-vanished opponents were still in the audience. Phagocytes, he proclaimed, were going to be increasingly useful to physicians—a prophetic statement, as we know today. "The phagocyte theory, founded more than a quarter century ago, was for many years vigorously attacked from all sides," he said. "Only lately has it been accepted by a large number of scientists from all countries, and only since yesterday, so to speak, has it been applied in practice. There is all reason to hope, therefore, that in the future, medicine will invent even more ways of using phagocytosis to promote health."

From Stockholm, the Metchnikoffs traveled east by steamboat and train to an even colder and windier St. Petersburg. "Yesterday it even snowed a little," Metchnikoff reported to Marie and Émile from Russia's capital. "One can tell it's spring only by the occasional straw hats worn by the ladies."

Nearly forty years earlier, he had fled the chill of St. Petersburg in an attempt to save his tubercular wife. At the time, he was struggling for a place under the sun in Russia's academe. Now he was returning to a hero's welcome. The St. Petersburg Gorodskaya Duma, the city council, named a municipal lab after him and created a bacteriology prize in his name. Hailed everywhere as the "pride and glory of Russia," he was awarded a doctorate *honoris causa* by the Military Medicine Academy and invited to tour the city's hospitals and its central "disinfection chamber," which sterilized the possessions of people with infectious diseases. Newspapers bemoaned the circumstances under which as a "liberal young professor" he had once been forced to quit Russia's university system. More than a dozen Kharkov University alumni held a festive dinner for him, serving lamb, hazel hen cutlets, and asparagus.

The celebrations peaked when twenty-five hundred people packed the stalls, balconies, and passageways of the Duma hall to attend a gathering convened in Metchnikoff's honor by the local medical societies. ("I'm very shy before such a large crowd," he confessed later in a letter to Marie and Émile.) Honorary chairman Ivan Pavlov opened the meeting effusively, "We all in Russia have been feeling gloomy in the wake of foreign and domestic upheavals, but your appearance, Ilya Ilyich, has uplifted our spirit. We jointly welcome a gigantic Russian scientific force recognized by the entire world."

Greeted by a standing ovation, Metchnikoff delivered a lecture on cholera. As always, he stressed the need for preventive measures—a burning topic in St. Petersburg, recently hit by a cholera epidemic. "None of us can protect ourselves from the most common running nose, but we *can* protect ourselves against cholera," he was reported as saying in an article in *Birzhevye Vedomosti* titled A PROPHET IN HIS OWN LAND.

38

TWO MONARCHS

EARLY ON A CLOUDY MORNING in late May 1909, a bow-tied Metchnikoff in a gray frock coat and Olga—wearing a long A-line skirt, a white blouse, and a straw hat with a light-blue veil—got off the train at a dirty railway halt after an all-night ride south from Moscow. As it had rained hard the night before, the air was exceptionally fragrant and fresh. A troika that Count Leo Tolstoy sent drove them along a rutted road through forests and fields, until passing between two old white columns to enter Tolstoy's estate of Yasnaya Polyana. Here was an alley of vintage birch trees leading to a white two-story house, drowned in greenery and blooming lilac bushes, in front of which peasant women were rapidly strewing the ground with sand.

Close on the troika's heels followed an entourage of journalists and photographers who had been covering Metchnikoff's every move in Russia. His dramatic visit to Tolstoy symbolized an encounter between two conflicting worldviews: rationality and faith. "When it was learned that the great student of the human body had decided upon a pilgrimage to the great student of the human heart and soul, Tolstoy, the press was occupied almost exclusively with the meeting of the two monarchs of universal literature and science," reported the *New York Times* in its article METCHNIKOFF—"THE APOSTLE OF OPTIMISM"—ON THE SCIENCE OF LIVING.

Pilgrimage was indeed an appropriate word. Even though several years earlier Tolstoy had been excommunicated by the Russian Orthodox Church for his criticism of institutionalized religion, he was widely venerated as a moral and spiritual authority. He too had come to see himself more as a preacher than a writer. He believed *War and Peace* and *Anna Karenina* had been mere teasers for getting people to read his more significant works, such as *The Kingdom of God Is Within You*, in which he called for love of humanity and nonviolent resistance to evil. Despite his admonition that people should live by their own conscience, not his, he was besieged by visitors, just as Metchnikoff was in his Parisian lab. They flocked to Yasnaya Polyana from all over Russia and even from abroad to talk philosophy, vent grievances, or ask for guidance or money.

Years earlier, Tolstoy had been acquainted with Metchnikoff's older brother Ivan Ilyich, a magistrate in a regional court, whose death of stomach cancer at forty-five inspired him to write one of his most famous stories. It is common knowledge what Tolstoy had to say about the fictional Ivan Ilyich, probably a composite portrait, as was the case with many of his protagonists. Less well known is what real-life Ivan Ilyich had to say about Tolstoy. He had once told his younger brother Elie (Ilya Ilyich) that Tolstoy, for all his genius, was not much of a rational thinker. "For instance, as a professor of zoology, you are really well familiar with the science of forest wildlife. But if I go hunting, I wouldn't take you along to help me find a woodcock in the forest, I'd rather take a dog that will track it down much better than you, using only its senses," he said. "It's the same with Tolstoy. His intuition regarding the innermost content of the human soul is extraordinary, he can discern the most secret motives with amazing precision. But when a question needs to be resolved with reasoning and logic, Tolstoy's mind very often doesn't hold up to criticism."

It was precisely Tolstoy's knowledge of the human soul that captivated Metchnikoff. If anybody had the answers to questions about life and death, it would be Tolstoy, who in Metchnikoff's opinion had described the fear of death better than any other writer.

Because of this admiration, Metchnikoff had frequently addressed Tolstoy in his own writings long before they met, contesting his views on science in his earlier essay "Law of Life" and later in *The Nature of Man*. He never learned that Tolstoy, to whom he had sent a copy of the book, dispensed unflattering comments about his work, telling a relative that the book was "very interesting in its learned stupidity" and jotting down sarcastically in his diary in July 1904, "[Metchnikoff] has just now noticed that old age and death are not all that pleasant. Great minds, not such children of thought as you, *gospodin* Metchnikoff, have pondered how to make aging and death harmless, solving this question intelligently: by seeking the answer in the spiritual essence of man rather than in his buttocks."

But if Metchnikoff had known how much antagonism he had provoked, his desire to meet Tolstoy might have only grown. After all, having rushed to confront Robert Koch in Berlin some two decades earlier, he had a phagocyte-like predilection for hastening into hostile territory. "I had long wanted to get to know Tolstoy closer, learning in person what he really thought about universal issues that had fascinated me since my youth, especially the basis of morality, the meaning of life, and the inevitability of its end," Metchnikoff wrote later in an essay, "A Day with Tolstoy in Yasnaya Polyana."

Entering Tolstoy's house after being welcomed by one of his daughters, Metchnikoff and Olga noted the simplicity of its interior. The furniture was old and functional, with no attempt at luxury or elegance. Tolstoy, eighty, quickly coming downstairs in a loose white blouse belted below the waist, struck them with his vigor and youthful appearance despite his long white beard. "He stared us piercingly in the eye as if he wanted to see through us, but such a kind, gentle smile lit up his face that we were immediately relieved," Olga wrote later. Tolstoy greeted them with a remark of a novelist: "You look alike; it happens when people live together well and for a long time."

It was only after the guests had been given a tour of the garden and village and been served lunch with his family that Tolstoy declared he wanted to talk to Metchnikoff tête-à-tête. Suggesting that they visit neighbors, he invited Metchnikoff to ride together in a cabriolet he himself would drive, while Olga and others traveled separately.

As soon as the cabriolet passed outside the gate, Tolstoy tried to assure Metchnikoff that contrary to prevalent opinion, he was not at all hostile to science. "I highly value genuine science, one that is interested in man, his fate and happiness, but I'm an enemy of pseudoscience, which imagines it's done something uncommonly useful if it's determined the weight of Saturn's satellites or something of the sort," he said. Metchnikoff recalled in his essay that in return, he argued that the most impractical scientific investigations had sometimes done the greatest good to humanity; that the study of microbes, for instance, had begun long before their role in disease had been discovered. But he failed to catch on to the real reason Tolstoy had wanted to talk to him alone. Tolstoy later complained to a friend about his attempt to discuss religion with Metchnikoff: "I tried it for a bit, then fell silent and just let him talk. He believes in his science as if it were the Holy Scripture. Religious and moral questions stemming simply from a sentiment of morality are entirely foreign to him."

As they sat down to tea at the neighbor's, at a white-clothed table with a samovar, Tolstoy brought up the outrages of social inequality, a subject that tormented him. "It's so difficult to have servants," he said. "Sometimes, an old servant attends to a young man." He then lamented that after serving their masters a gourmet meal, the domestics had to eat leftovers or poor-quality food. Metchnikoff immediately suggested that science could offer a solution. Once, he said, he'd been asked to inspect an opulent mansion of a bourgeois family near Paris whose members kept getting ill. He had quickly discovered that their servants lived in such unhygienic conditions that, lacking a toilet, they spilled their excrements into the garden, polluting the vegetables used to prepare the masters' meals. "No wonder you are getting sick, you are eating the filths of your servants," Metchnikoff had told the mansion owners. This example, he announced, showed that social progress—here, the well-being of servants—could rest on such selfish motives as the masters' concern about their own health. "Progress doesn't necessarily have to be based on people's love for one another," he said. He did not notice that Tolstoy, by his later admission to a friend, was shocked by this scatological approach to moral values.

When humane treatment of animals came up in the conversation, prompting Tolstoy to speak of his current aversion to hunting, even though he had enjoyed it greatly in his youth, Metchnikoff countered that wild animals were likely to suffer less if killed by a hunter's bullet than if they fell prey to predators, as inevitably happens in nature when they grow old. Besides, he added, a ban on hunting would lead to a proliferation of predators, which would cause harm to people. Tolstoy, a vegetarian, strongly disagreed. "If we subject everything to reasoning, we might arrive at the most unbelievable absurdities," he said. "We may even end up justifying cannibalism." Metchnikoff wrote later in his essay, "This response seemed to me the most remarkable statement of our entire conversation—it revealed all of Tolstoy. . . . His sensitivity and impressionability had overtaken his artistic nature, so that reasoning and logic had moved to the background."

Tolstoy further declared he had no special interest in longevity. "The trouble is not that our life is too short but that we live badly, contrary to our own conscience," he said. When Metchnikoff went on to describe the loves of Hugo, Ibsen, and other famous elderly men, Tolstoy stated curtly, "I won't supply you with any material in this respect." In his youth, he had given free rein to his powerful libido, but since turning to religion in his older age, he had started seeing sex as so sinful he felt guilty even when desiring his own wife.

When returning to Yasnaya Polyana, Tolstoy rode on horseback, alone.

Back at the house, upstairs in his study, Tolstoy asked Metchnikoff, "Tell me, why, in fact, have you come here?" Metchnikoff felt understandably taken aback. "Somewhat embarrassed, I told him I'd wanted to learn more about his objections to science and express my deep respect for his works of fiction, which I considered incomparably superior to his philosophical works," he wrote in the essay.

In the evening, after everyone listened to a visiting professor of music play Chopin, a rare moment of harmony was reached in the conversation. "In science, it's much more difficult to create something new than to destroy what already exists," Metchnikoff said. "That's true in everything, especially in philosophy," said Tolstoy. The one other thing they agreed upon was yogurt. Even though this probably never came up during the visit, in an earlier letter to his wife, Tolstoy had mentioned he believed he was feeling well because of the sour milk he was drinking, as recommended by Metchnikoff.

When Metchnikoff rose to catch a night train back to Moscow with Olga, Tolstoy announced, "I'm glad that despite our totally different starting points, in the long run we strive for a common goal: seeking the good of people!" It was a diplomatic way of acknowledging that neither of them had budged from their original positions. "I'll try to live a hundred years in order to please you," he added with a laugh.

Tolstoy did not keep the promise. About a year and a half later, he was to secretly escape from home accompanied by his physician, running away from the comfort of his estate, the unbearable flux of visitors, and a wife who resisted his attempts to give away all their property. While on the run, he contracted pneumonia and died.

Metchnikoff remained blissfully unaware of Tolstoy's remarks about him. After his and Olga's visit, Tolstoy had told a friend about Metchnikoff, "He's a nice, simple person, but science is his weakness, just like some people have a drinking problem." He evidently couldn't forgive Metchnikoff for encroaching, with his orthobiosis, on the core of his own teachings about the meaning of life. Tolstoy later added, "How many types of flies do you think scientists have counted? Seven thousand! Who can then find the time for spiritual questions?"

In correspondence with a scientist friend after Tolstoy had died, Olga addressed the friend's question—had Tolstoy indeed conquered his fear of death through religious faith, as he had claimed? "He even claimed to have desired death," Olga wrote. "But his wife and children said that deep down, he hadn't overcome this fear because whenever he fell ill, he grew anxious and consulted physicians." With a scientist's precision, Olga commented, "Perhaps he was afraid of illness and not of death."

In the Yasnaya Polyana essay, written two years after Tolstoy's death, Metchnikoff, for his part, offered an explanation for the writer's hostility toward science. A positivist, materialist worldview ran counter to Tolstoy's religious beliefs, he wrote. Moreover, science could not help him cope with his fear of death. Therefore, Metchnikoff argued, Tolstoy had begun to question scientific achievements, once even telling a physician friend he doubted the validity of Pasteur's research on rabies. "This conversation took place in late 1894, that is, at a time when the value of Pasteur's antirabies vaccine had been recognized even in Germany," Metchnikoff wrote. "Tolstoy's self-confidence in passing judgments on issues totally unfamiliar to him prevented him from seeing the truth."

For Metchnikoff, of course, the "truth" was science. It never occurred to him that not all questions could be meaningfully addressed by scientific research. And he was sure only science could one day bring about a perfect harmony between sentiment and reason.

39

A METCHNIKOFF COW

"A SURGICAL MONSTROSITY" WAS HOW George Bernard Shaw referred to a reckless procedure inspired by Metchnikoff's theories: removal of the colon. Raising a storm of criticism among physicians as well, the surgery was performed by one of Britain's most talented and controversial surgeons.

Sir William Arbuthnot Lane was a tall, thin man with an imperturbable demeanor and the dexterity of a born mechanic. People came from all over the world to undergo his various innovative surgeries. Among his patients were members of the royal family. Like many Victorians, he was intensely preoccupied with his own and other people's bowels. Reading *The Nature of Man* and then meeting Metchnikoff himself in 1904 affected him deeply. An evolutionist interested in vestiges of the past in the human body, Lane decided that the entire colon was a "vestige" and a "cesspool."

Unlike Metchnikoff, Lane did not necessarily believe the human organism was inherently disharmonious. Rather, he blamed modern civilization. It had forced humans into "the habit of keeping the trunk constantly erect," he wrote, which, in turn, impelled the large intestine to form kinks and bends, making it difficult to empty its contents and thereby augmenting "autointoxication." He began to treat severe chronic constipation, which in extreme cases can be life-threatening (for a while it came to be known as Lane's disease), by slashing the patients' abdomens wide open, untangling their intestines, and cutting out all the five feet or so of the colon.

In 1913 Lane was finally to hold what a biographer of his called a "Great Debate" on his surgeries at the Royal Society of Medicine in London. His opponents, in contrast to Metchnikoff, believed the human organism was fundamentally harmonious. One Professor Arthur Keith argued in defense of the colon, "It is hard to believe that a great structure which has served that long chain of ancestors, carrying man's lineage through the secondary and tertiary periods of the earth's formation and assisting man to become the dominant and universal species of the world, should suddenly fail him." Lane lost the debate and later switched to preventive medicine, encouraging people to avoid constipation through diet and exercise, but by then he had shortcut the bowels of dozens of patients.

Metchnikoff himself never recommended a preventive removal of the colon—just the opposite. In his 1904 "Old Age" lecture, he had warned that such a move "could not be discussed seriously." But when Lane began to perform the operations to treat constipation, Metchnikoff never protested or tried to encourage him to resort to less radical means. In fact, he was pleased to hear people managed just fine without their colons, checking up on Lane's patients whenever he was in England and asking for one of these patients to be sent to the Pasteur Institute for observation. He did, however, call for caution "in the present state of surgery." "Dr. Lane, an unusually skillful and courageous English surgeon, has dared to resort to an operation instead of an ineffective internal treatment," he said in a 1909 public lecture in Stuttgart. "This surgery is naturally very dangerous and at present leads to numerous deaths. Dr. Lane has already performed more than fifty such operations and last year described thirty-nine in detail. He lost nine patients, or 23 percent—a significant mortality rate—whereas the remaining thirty patients benefited greatly from the intervention." Metchnikoff then noted hopefully that "when surgical technique improves, mortality will drop."

No less controversial than Lane's surgeries was the ongoing hysteria over the putative role of sour milk in longevity. Even though the issue continued to be hotly debated at dinner tables around the world, it came under severe criticism from experts. *Foods and Their Adulteration*, an authoritative book published in Philadelphia, added to its 1907 edition a new section, "Sour Milk and Longevity," in which the author, Dr. Harvey W. Wiley, tried to dispel yogurt's longevity mystique. Excessive claims, he wrote, "only serve to bring the whole subject of the use of sour milk into deserved contempt." Voices of reason were heard in the medical press too. The American magazine *Medical News* noted that "Professor Metchnikoff, of the Paris Pasteur Institute, is distinctly what the French call *un homme sérieux*, otherwise one would be tempted to suspect that he amuses himself by practicing on the gullibility of the public."

Metchnikoff himself tried to temper the grandiose claims. "Contrary to what many journalists have made me say, I have never, in any of my publications on the subject, asserted that curdled milk is able to prolong life," he pleaded in a popular article in 1909, "The Utility of Lactic Microbes." But the easy recipe for longevity was too alluring to be abandoned quickly.

Fanning the fad, journalists sometimes invented quotes to add spice to a story. This is obviously what happened in the summer of 1912, when the *New York American*, in the full-page article PROFESSOR METCHNIKOFF DISCUSSES YOUTH AT 100 YEARS, "quoted" Metchnikoff as saying, "With the use of my preparations, an age of one hundred and fifty years will be nothing unusual in the near future. A woman at the age of sixty will not be then called an old maid. She will be able to marry even at the age of ninety or above a hundred."

Metchnikoff must have been greatly embarrassed to receive in the mail another newspaper clipping, sent by an American physician, who had written on the margins in red pencil: "I sincerely hope you are not responsible for this ridiculous, not to say insane 'slush.'" The abundantly illustrated full-page article from the Chicago broadsheet *Sunday Record-Herald* quoted Metchnikoff as saying the cow's mouth was "a veritable hotbed of microbes," and was titled accordingly: "BRUSH THE COW'S TEETH AND LIVE 250 YEARS," SAYS PROF. METCHNIKOFF.

As was to be expected, the sour-milk fad inspired satire, especially in the United Kingdom. Making fun of Metchnikoff's mention of Old Parr, the seventeenth-century Englishman who had reportedly lived to 152 years, the *Daily Chronicle* mockingly cited "evidence" for the sour-milk theory of longevity in an old poem, "Life of Old Parr," that mentioned green cheese and butter-milk.

The fad quickly became embedded into popular culture. Perhaps the epitome was its featuring in the pantomime *Jack and the Beanstalk,* a spoof on the fairy tale, presented at the Drury Lane theater in London in December 1910. In the show, the king is prescribed the "sour-milk cure" for his gout and visits a sour milk farm. But contrary to promises, Priscilla, a cow belonging to Jack's mother, gives sweet milk instead of sour and is sold for a bag of beans. In a raving review, the *Times* of London listed "a Metchnikoff cow" among the show's greatest attractions.

The joking and the jeers never swayed Metchnikoff away from his views on longevity. His best supportive evidence was himself. "I'm sixty-four and you see how well I'm preserved," he bragged to a journalist on his visit to Russia. "I feel much better now than I did at thirty-five, especially since I've started eating Bulgarian sour milk three times a day." And his youthful energy certainly never failed him when it came to launching imaginative research projects. At sixty-six, accompanied by Olga and five colleagues, he led an expedition to Russia's Kalmyk steppe, sailing for five days along the majestic expanses of the Volga, then crossing the desert for three months to study tuberculosis and the plague among the indigenous population.

Back in Paris he continued to study infectious diseases—supporting the notion, for example, that antityphus vaccinations were effective in humans—but his main focus was research into gut microbes. His goal was to test the hypothesis that these microbes indirectly contributed to aging by releasing toxins, which, in turn, overstimulated the phagocytes. Several volumes of the Pasteur Institute's *Annales* from those years contain numerous studies conducted on this topic under his guidance, including his own four papers titled "Studies of the Gut Flora," as well as those by his students and visiting scientists. He had always had numerous disciples, but now so many people wanted to train in his lab they had to sit two at a desk. Metchnikoff, a natural leader, circulated among them, dispensing instructions.

The "zoo" in his lab kept expanding with exotic new species. It already had snakes and a caiman, a small crocodile on which Metchnikoff tested the effects of tuberculosis and other disease germs. For the study of gut microbes, Metchnikoff ordered from India a shipment of several dozen large bats of the genus *Pteropus,* known as "flying foxes." These beady-eyed creatures, with a fox-like head and wide wing membranes, have almost no colon and therefore hardly any gut microbes. They offered the unique opportunity to test in mammals an important question, still debated today: are gut microbes indeed indispensable to the organism's well-being, as Pasteur had suggested? The bats did just fine when fed sterile food, but the question proved hard to resolve. In other animals, the microbes did seem to be needed for proper growth. For example, in experiments Olga conducted, tadpoles grown in sterile conditions developed much slower than the ones grown in a regular, germ-infested environment.

To check whether gut microbes affected body chemistry, Metchnikoff regularly examined his own urine, noting whether test results changed after he had consumed lactic acid microbes. In parallel, he fed several groups of rats with meat, plant food, or a mixture of the two. He found that the chemical composition of the gut varied from one animal to another even when the rats were fed identical diets, suggesting that response to food was indeed altered by the gut microbes. Metchnikoff hypothesized that in humans, too, response to food was affected by these microbes—a daring idea but one that is now, a hundred years later, receiving resounding support.

What, then, is the optimal composition of the gut flora? Metchnikoff argued that just as some forest mushrooms are edible whereas others are poisonous, so too can gut microbes be useful or dangerous. "The idea is to define precisely the two categories, and to engage the beneficial bacteria in fighting the harmful ones," he wrote. He also argued that the gut flora could be altered by diet and that this flora changed in the course of a person's life; he observed, for example, that it was richer in adults than in babies.

Remembering that his ultimate interest was aging, Metchnikoff, aided by his students, experimented on rabbits, guinea pigs, rats, mice, and macaques to study the effect of gut microbes on the hallmark symptom of aging: hardening of the arteries. To stress that this effect was real, he often referred to an Argentinean study in which calves suffering from a certain intestinal infection were found to develop a severe hardening of the arteries.

Old age wasn't his sole focus, however. His fourth and last paper on gut microbes, "Diarrhea in Infants," published in 1914, was devoted to this major cause of child mortality at the time. A prevailing view among doctors was that most cases were not triggered by microbes but by food or summer heat. Metchnikoff, after feeding newborn rabbits with excrements of sick babies, proved that infant diarrhea was for the most part microbial in origin, calling for "immediate hygienic measures."

Among scientists, Metchnikoff's studies of the gut flora provoked extreme reactions, both of the positive and negative sort. On the positive end were scientists who picked up the topic themselves. Among these were Yale University bacteriologists Larry Rettger and Harry Cheplin. Without focusing on aging, or on phagocytes for that matter, they launched extensive investigations into lactic acid microbes, publishing a book, *A Treatise on the Transformation of the Intestinal Flora,* in 1921.

On the negative extreme were dismissal and mockery. German pathologist Hugo Ribbert, for instance, thought that Metchnikoff's linkage of gut microbes to aging was a "futile fantasy." Another testimony to negative reactions comes from the memoirs of a bacteriologist who for a while worked as a research assistant at the Pasteur Institute. He recalls that because Metchnikoff was forever analyzing the contents of intestines, some people, apparently even other scientists, nicknamed him "*Fouille-Merde,*" "Shit-Digger," behind his back.

What certainly did not agree with many researchers was Metchnikoff's media stardom. As a distinguished scientist turning out provocatively appealing theories, he was a household name around the world. In 1911, when the British *Strand Magazine* polled politicians, scientists, and writers on who they thought were "the ten greatest men now alive," Metchnikoff was placed ninth on a list that started with, in descending order, Thomas Edison, Rudyard Kipling, Theodore Roosevelt, Guglielmo Marconi, and Lord Lister. A commentary in the *British Medical Journal* seeped with disapproval. "The list on the whole is not so foolish as might have been expected," the *Journal* wrote, extolling the accomplishments of Lord Lister. It then added a nasty remark obviously aimed at Metchnikoff, the only other medical researcher on the list, nominated by many voters together with Lister: "Of several of those who are placed by [Lister's] side among the elect, it might be said with truth that their greatness was more in the noise they have made than in any good they have done."

40

RATIONAL WORLDVIEW

In the first decade of the twentieth century, the nationalist mood in France had waxed and waned, but as the century progressed, disputes with Germany over African colonies reawakened anti-German sentiment and the desire for revenge. French playwrights produced a spate of patriotic plays, books exalting bravery became bestsellers, and lecturers hardly dared mention German methods or ideas for fear of catcalls. At the end of 1911, when France and Germany signed an agreement on the colonies—denounced in both countries as a humiliating diplomatic defeat—French nationalism, according to historian Eugen Weber, crescendoed into "a widespread, chauvinistic feeling." Preparedness for war with Germany, which now seemed inevitable, became France's urgent concern, not only in building military strength but in forging national unity—an ambience fostering prejudice against everything perceived as ill-assimilated or foreign.

Olga describes an atmosphere of hostility in which Metchnikoff became a target of xenophobia:

> Foreigners were accused of inundating France, occupying jobs, and aggravating the struggle for existence. At first these were but vague allusions, but little by little, the nationalist circles crossed all bounds of justice and decency; the attacks turned into brutal provocations. The contemptuous term *métèque* [a

> pejorative for immigrants] was revived. One nationalist newspaper led a particularly vicious propaganda, balking at no means to slander its victims, Metchnikoff among them.

As nationalism was reaching its peak, Metchnikoff's affiliation with the yogurt company Société Le Ferment suddenly triggered the fury of the French press. The daily *Le XIXe Siècle* declared on its front page in June 1912, "If the profits from these sales go into the institute's budget, serving the progress of microbiological science, that's fine. But if these novelties of uncertain efficacy are exploited for someone's personal benefit, then that's truly abusive."

Some of the critiques were outright chauvinist. "For a few years, under the influence of Metchnikoff, who keeps repeating—surely in the hope of persuading himself—that it's easy to push off aging and postpone death indefinitely (this attitude goes under 'optimism')—people have been introducing numerous microbes into their digestive tracts," wrote the *Journal des Débats Politiques et Littéraires.* Proposing to launch a quality control service for antiaging products at the Pasteur Institute, the paper added, "The public would be pleased should such a service, created in an institution founded with a national subscription, be confined to the French rather than to foreigners."

The "particularly vicious" attacks Olga mentioned were probably the ones in *L'Action Française,* which Weber described as a daily "on the lunatic fringe of the Right." Metchnikoff had recently announced that in the intestines of a dog, of all places, he had found a microbe he called *glycobacter,* which supposedly could turn starch into sugar that was needed for the production of acids in the intestines. The thought of the microbe's recent place of residence didn't repulse Metchnikoff or his students in the least. For a while, some lab members not only fed potatoes seasoned with a pure culture of the microbe to rats and macaques in the lab but also ate such potatoes themselves to study glycobacter's effects. This line of research was later to prove a dead end, but long before this was found to be the case, *L'Action Française* mocked it in venomous assaults, spiced with an anti-Semitism that brings to mind Nazi pamphlets of later years. In July 1912, in a front-page article titled THE WONDERFUL REMEDIES OF *PÈRE METCHNICROTTE*

[Papa Metchnidroppings], it accused Metchnikoff, "the Jew of dog droppings and of the Pasteur Institute," of launching dubious "elixirs of longevity" and profiteering through Le Ferment.

Only letters to editors from Metchnikoff's colleagues at the Pasteur Institute and elsewhere finally put an end to the ugly campaign. In a rebuke to the *Journal des Débats*, Roux, as the institute's director, expressed his highest esteem "for my friend Metchnikoff, who for twenty-four years had devoted himself to the Pasteur Institute with admirable disinterestedness."

The nationalist upswing was part of a tide that would soon engulf a large part of the world in a bloody conflict. Metchnikoff was much more concerned about the accompanying backlash against science and rationalism. In the summer of 1912, he prepared for publication his *Forty Years of Searching for a Rational Worldview*, a Russian-language collection of philosophical essays he had authored from his midtwenties to his midsixties. Its title suggests it was intended as an affirmation of rational thinking.

The book was to become his swan song. At the end of the preface, Metchnikoff reasserted his utopian vision of the distant future. "With time, when science eliminates misery, when it will be possible not to agonize over the health and well-being of loved ones and human life will run its normal course, people will be able to rise to a higher level than now, devoting themselves with greater ease to serving the loftiest goals."

But in the more immediate future, the world was about to fall into shambles.

41

THE LAST WAR

IN THE SUMMER OF 1914 Metchnikoff and Olga rented a cottage in Saint-Léger-en-Yvelines, in the heart of a forest southwest of Paris. "Fields with an endless horizon, forests casting shadows over luxurious ferns, carpets of moss and heather of all hues, mysterious ponds," was how Olga described the scenery. Metchnikoff had little taste for sightseeing or travel for travel's sake—declining, for instance, numerous invitations to visit the United States. But he relished spending time in nature. To his forest summer residence, he brought with him boxes containing cocoons of white-haired silkworm moths.

He believed the moths could finally provide him with a model for studying his proffered "natural death," one that is not caused by accident or disease but "depends on the organism itself, its most inner essence." These insects live for about a month until they die of internally predetermined causes, which, he thought, could hence count as natural. As the moths began to hatch, all the tables and mantelpieces in the living room were covered with their white flakes. This research, to appear in the October 1915 issue of the *Annales*, was to be Metchnikoff's last published study.

His search for a "natural" death on the eve of an armed conflict about to result in millions of utterly unnatural deaths now seems ironic, as are his upbeat statements on continued peace in Europe. "Appeals for peace are now heard even from world leaders who only recently rattled their sabers every time they spoke in public," he wrote in one newspaper article. "As a result, human life is valued ever more highly."

But his visions of a brighter future were not entirely unfounded. After all, the continent had managed to avoid major warfare for decades. Until the outbreak of hostilities later to be known as World War I, Metchnikoff was convinced the progress of civilization would make violent conflicts disappear.

From the time he was a child he had abhorred the use of brute force. The first vivid memory of his life was of himself as a terrified five-year-old clinging to his mother's skirt while he watched a company of hussars pitilessly beat up insurgent muzhiks, throwing one peasant woman to the ground, earth forced into her mouth. "This episode marked him for life, filling him with a hatred of all savagery and violence," Olga wrote.

Then, on July 28, Austria-Hungary declared war on Serbia. In a deadly domino effect, the conflict quickly involved Austria's ally, Germany, which in turn pitted itself against Russia and its allies, France and the United Kingdom. Metchnikoff refused to believe war had actually started. But, in the words of a British statesman, the lamps were already going out all over Europe.

On the stormy night of August 1, woken up by a pounding on the gates of a nearby house, the Metchnikoffs, in a flash of the lightning, saw horsemen with lanterns. They had come to deliver conscription notices. Two days later, Germany declared war on France.

Rushing back from Saint-Léger-en-Yvelines to Sèvres, Metchnikoff then hurried to the Pasteur Institute. When he came back in the evening, Olga was in shock. "I'll never forget his return home," she wrote. "Waiting for him as usual by the gate of the station, I didn't immediately recognize him from afar. I saw an old man, stooping as if under a heavy burden. His habitual vivacity had given way to a profound depression. He told me in a broken voice that the institute was under the orders of the military authorities, completely disorganized for scientific work. The young had been drafted, the laboratories were empty. Lab animals, even the apes, had been killed, for fear of a siege of Paris and a lack of food to feed them."

Very soon, the entire Pasteur Institute was recruited into the war effort—both its premises and most of the remaining personnel. (More than half of the 230 staff members had been mobilized.) Shacks were set up on campus and makeshift facilities were added in the institute's corridors to help produce medicines and vaccines against dysentery, meningitis, and other infections that plague battling troops. Instead of research animals, the institute's kennels now housed cows that gave milk for hospitals and children's homes. In the first few weeks of the war, the Pasteur Institute supplied 670,000 doses of the antityphus vaccine, ultimately producing millions. During one German offensive, the institute manufactured antitetanus serum at the rate of twenty thousand ampules per day, which gives an idea about the scale of the bloodbath.

As the German armies approached, panic seized Paris; thousands of Parisians fled west. The government moved to Bordeaux. German Taube planes regularly circled over the city, once dropping a bomb near the train station just as Metchnikoff and Olga were leaving it. At night, the sky was crisscrossed by the gigantic beams of searchlights, and cannon fire thundered in the distance. "All that's left in Paris," one journalist wrote, "is Paris, its monuments, squares, its hills, its river—and what a marvel, deserted Paris, the great boutiques closed, a rue de la Paix, a rue Royale, an avenue de l'Opera traveled only by authorized military vehicles."

Metchnikoff decided to stay in Paris to assist Roux at the Pasteur Institute. Olga worked in a surgical ambulance after taking a nursing course. In his deserted lab, Metchnikoff wrote the book *Founders of Modern Medicine: Pasteur, Lister, Koch.* His goal was to affirm the value of science through the concrete examples of these three great men. "Let us hope that this unparalleled carnage for a long time discourages people from fighting," he wrote in the preface. He emphasized this wish for peace in a note to his godchild Elie, Lili's little brother, writing, "This war will certainly be the last one."

This "last" war broke Metchnikoff's spirit. His heart, which had already started giving him trouble in the preceding year, began to fail in the wake of his tendency (surely not uncommon) to fall physically ill in periods of heightened anxiety or grief. One attack of tachycardia, a rapid irregular heartbeat, occurred in the fall of 1914, another in the spring of 1915.

From a youthful septuagenarian, he suddenly turned into a bent, slow, and listless old man. "The war cut him down like an oak," Olga was to recall years later in her diary. In her book, she wrote that little children in the street, for whom he always carried sweets in his pockets, started calling him Père Noël, "Father Christmas."

42

A TRUE BENEFACTOR

In May 1915, when friends and colleagues gathered in the Pasteur Institute library to celebrate Metchnikoff's seventieth birthday, he accompanied his own address with a chart showing that he came from a short-lived family. His brothers had died in their forties or fifties, his parents and sister in their mid to late sixties. Considering his heredity, he was lucky to have reached seventy, he declared.

He stressed this on every occasion, adding more explanations. "One shouldn't forget that I started living very early (having published my first scientific paper at eighteen), and spent my entire life in great anxiety, in full swing," he wrote in a diary of self-observations he had started keeping at the first signs of heart trouble. "The polemics over phagocytes might well have killed or completely weakened me much earlier. . . . Moreover, I started following what I consider a rational hygiene only after reaching the age of fifty-three, when I already had the signs of arteriosclerosis." He was tormented by the thought that his death would deliver a fatal blow to his theories on aging. "Let all those who expected me to live one hundred years or longer 'forgive' me my premature death in light of the noted above circumstances," he wrote.

The urge to overcome the fear of death preoccupied him constantly. Some four years earlier, he had admitted to a visitor from Russia that he hadn't managed to conquer this fear. On his seventieth birthday he proclaimed in his diary, "Now I'm positively unafraid of death." But his endless dwelling on the topic suggests just the contrary. Rather than disappearing, his fear seems to have morphed into a persistent anxiety, mainly about the well-being of his intimates. "The worry over the happiness and health of my loved ones keeps getting stronger, I don't know how I could bear it in the past," he wrote in the same entry.

In early December 1915, after Metchnikoff had caught a cold, his health deteriorated quickly. His heart kept weakening and he had difficulty breathing, with occasional attacks of wheezing or coughing, sometimes with blood. Nights were the worst. Having barely fallen asleep, he woke up in anguish, drenched in sweat. He and Olga moved into a small apartment on the upper floor of the Pasteur Institute hospital, where Metchnikoff could get medical care and receive endless visitors even when bedridden—friends, students, colleagues, Russian journalists, and politicians, all eager to see Papa Metch. "Throughout his illness, his warmth and gentleness were touching, which didn't prevent him from voicing his opinions with his usual directness, but I noticed that this no longer offended people," Olga wrote in her book.

On May 16, 1916, a day after his seventy-first birthday, Metchnikoff jotted down in pencil in his diary with a shaking hand, "My dream to die quickly, without a long illness, has not come true. I've been bedridden for over five months." To his disappointment, he experienced no sign of a death instinct whatsoever. "I think that my desire to get well and continue living is partly due to practical circumstances," he wrote apologetically. "War has undermined our income from Russia. My wife, being so impractical, might face dire straits in the case of my death." It was the last diary entry written in his own hand.

On June 18 he dictated to Olga one last lengthy diary entry. It lists observations in defense of his theories. His poor heredity explains his approaching premature death; his own "life instinct" had weakened in recent years, just as he had predicted; yet he had managed to stay active in research till a later age than all bacteriologists of his generation. All this "can serve as proof that my orthobiosis had really attained the desired limit." Till the very end, the fate of his theories troubled him no less than his own.

In late June, at Roux's suggestion, Metchnikoff and Olga moved from the hospital, where the rooms had become hot and stuffy, into the airier former quarters of Louis Pasteur on the upper floor of the institute's northern wing. The ornately furnished apartment had a piano on which Olga played Metchnikoff his favorite music. What delighted him most was being close to his laboratory, which was across a gallery on the upper floor of the institute's other wing. From time to time, he still nurtured the hope that he would return there one day.

By now he could barely breathe without oxygen or sleep without opiate injections. His heart was weakening further, and the tachycardia was constant. He had congestion of the right lung and fluids in the lung's lining. His cough kept worsening, and he was spitting blood.

One day, toward mid-July, he told Doctor Alexandre Salimbeni, a former student of his who had accompanied him on his expedition to the Kalmyk steppe, "*Donc alors*, this is my last week. You'll be the one to perform my autopsy." Maintaining his playfulness on his deathbed, Metchnikoff added, "I can see you, with your head bent to the side as you do when you are preoccupied, looking a little sad. Don't be. All this has little importance." He was probably trying to soothe himself as much as his former student.

On the afternoon of July 15, feeling suffocated, he asked Salimbeni, "You are a friend. Tell me—is this the end?" To the doctor's protests, Metchnikoff responded with instructions about his own autopsy: paying special attention to his intestines. "I beg you, don't make such abrupt movements, you know it's bad for you," Olga pleaded with him after he'd moved brusquely. He did not answer.

"I looked up," she wrote in her book. "His head was thrown back on the pillow, his face had turned bluish, his eyes had rolled behind half-closed lids . . . Not a word, not a sound . . . All was over . . ."

On the cloudy morning of July 18, Metchnikoff's friends and students, as well as representatives of the French and Russian governments, filled the domed, temple-like columbarium of Paris's Père-Lachaise cemetery. After his coffin disappeared behind the incinerator's black curtain, they sat in complete silence as the cremation took place. In accordance with Metchnikoff's wishes and his allergy to pomposity, there were no speeches, religious rites, or wreaths. Olga, looking pale and worn, was in the front row next to Lili, who had just turned thirteen. Lili later recalled that throughout the hour-long wait, Olga held her hand, pressing it from time to time. When a car took them back to the Pasteur Institute after the ceremony, in its back was the small red-marble urn with Metchnikoff's ashes, making its way, as he had requested, to the institute he never wanted to leave.

Even though war reports dominated newspapers, Metchnikoff's death was major news around the world. Next to the headline 713TH DAY OF THE WAR, *Le Figaro* eulogized him profusely on its front page, writing with a touch of wartime patriotism, "He was one of ours, having devoted the most fruitful years of his admirable life to working among us and teaching and writing in our language, which he had mastered in all its delicious expressiveness. For that, France will remain forever moved and grateful."

International press was unanimous in praising Metchnikoff, ranking him as "one of the commanding old men of his field." The journal *Nature* called him "one of the most remarkable figures in the scientific world." The *British Medical Journal* announced that "the whole scientific world will mourn the death of [Metchnikoff], whose researches may truly be said to have changed the face of pathology."

When it came to Metchnikoff's work on aging, the tone became more cautious. The *Nation* reminded readers he had enough to his credit to sustain him through controversy: "However untenable some of the conclusions of Metchnikoff, however radical some of his beliefs, there is no doubt that his startling theories were one expression of the qualities that made him great." The influential New York weekly *Literary Digest* wrote about his dietary advice, "He harped on the theme so much his name came dangerously near being a joke. But his 'long-life' preachments were the least of his work."

Predictably, Metchnikoff's death was a blow to all followers of his theories of longevity, a sentiment echoed in many headlines. *Los Angeles Times*: METCHNIKOFF PASSES AWAY. FAILED TO ACHIEVE THE ALLOTTED CENTURY AND A HALF. *Chicago Daily Tribune*: ADVOCATE OF THE "SOUR MILK CURE" FOR OLD AGE SUCCUMBS TO HEART DISEASE. *Current Opinion*: PASSING OF THE GREAT SCIENTIST WHO SOUGHT A TRIUMPH OVER DEATH. The *Sunday Times* noted with characteristic understatement, "The death of Professor Metchnikoff, for whom the most knotty problems of science offered no difficulties, was in distinct disaccord with his longevity beliefs." The *Medical Times* was more blunt: "These cranky ideas which seem to germinate in the brains of otherwise extremely clever and acute men all end in the same way."

A few obituaries ventured guesses as to how Metchnikoff would survive in history. The *New York Times* predicted he would be remembered "as a discoverer and a true benefactor of the human race as well as the propounder of startling though amiable theories."

That was not to happen. Closest to the mark was the *Scientific American*. "It would be interesting to compile a list of the popular distortions of scientific reputations," the magazine wrote in its obituary, "The Dean of Bacteriologists," stating further, "Metchnikoff is condemned to be handed down to posterity as the man who advocated a diet of sour milk, the man who thought that people should live to a hundred and fifty and then died at 71 himself."

43

OLGA'S CRIME

In the winter of 1916, the first one after Metchnikoff died, during a spell of illness, despair, and solitude, Olga, fifty-eight, committed an act that felt to her like a crime—and in a way it was. She was entirely alone in Sèvres. All her friends had either been mobilized or had left Paris, fearing a siege. Roux was hospitalized after an operation. In the background was a constant thudding of bombs dropped by airplanes. Fuel for heating was so scarce she caught a cold and became very ill. She thought she wasn't going to make it.

Sitting on the floor before the fireplace in Metchnikoff's study, she reread the hundreds of letters she and Metchnikoff had preserved over the years, then threw them into the fire one by one. "I don't know if I actually cried, but I remember that I felt trembling and sobbing inside," she was to recall in her diary. "Now many years later, I can only think about that day of letter burning as murder."

She burned Metchnikoff's entire correspondence with his first wife, Ludmila. He had requested that it be destroyed after his death, calling it "too intimate." Even Olga's own letters to Metchnikoff went up in flames. "Only his letters to me I couldn't destroy: I wanted them to be burned together with myself, as I thought I was going to die soon," she wrote.

Those four hundred letters, as well as Olga herself, were in for an entire series of misadventures. Having recovered from her illness, Olga set out to fulfill her promise to Metchnikoff: completing his biography, which she had started writing while he was still alive. Published in French, Russian, English, and several other languages, it helped her make ends meet. The Soviet regime had expropriated her family's estate after the October 1917 revolution, depriving her of her main source of income. Struggling to survive on the Pasteur Institute pension, she sold her piano and, for a total of 30,000 francs (a considerable sum at the time), some sixty of her paintings and sculptures through an exhibition at the Artes gallery in Paris. She also sold the house in Sèvres, moving to a small structure on the Pasteur Institute campus dubbed *izba,* Russian for "peasant's hut." In the wake of the revolution, she was busy welcoming family members who had escaped from Russia and, with Roux's help, assisting Russian scientists seeking refuge in France.

Roux and she remained close, just as they had been when Metchnikoff was alive, but apparently not any closer. Roux's niece Mary Cressac argues in her uncle's biography that for years Roux, himself unmarried, was romantically involved with another married woman, Marie Delaître. His involvement with Olga, before and after Metchnikoff's death, came only second, Cressac wrote. "At least in appearance, Metchnikoff's passing changed nothing in the relationship between Olga Metchnikoff and Émile Roux," she wrote. "He continued to pay her the same discreet visits."

Olga hoped to organize a memorial room at the Pasteur Institute for Metchnikoff's papers and belongings, but in the disarray that followed World War I, her plan never materialized. Just then, Metchnikoff's former students in the Soviet Union pleaded with her to hand over his archive to Russia. In 1926 she took all his documents and various objects to Moscow, where she attended the opening of her husband's museum, which former students created in the Metchnikoff Moscow Bacteriology Institute just outside the city. She wrote to Lili, "All the portraits are assembled in the museum, which is like a secular chapel—you can't imagine what a cult of Elie they have in Russia!"

But Olga held on to the stack of Metchnikoff's letters to her that she had been unable to burn, packing them with her most precious belongings wherever she went. In 1933, at age seventy-five, after Roux died, she left the Pasteur Institute campus, living for a while in the seaside community of La Favière on the Côte d'Azur, where many White Russians had settled after escaping their homeland in the wake of the revolution. When the Germans occupied France in May 1940, Olga, fearing for the safety of Metchnikoff's letters, tried to send them to the Moscow museum, but this proved impossible in war-torn Europe. By the fall of 1943, when hostilities were expected to break out in the south of France, Olga, by then eighty-four and ill, returned to occupied Paris, leaving Metchnikoff's letters behind in a secret location.

Olga died on July 24, 1944, in the Pasteur Institute hospital, a month before French and American troops finally liberated Paris. Her last wishes reveal that till the end of her life, her loyalties were with Metchnikoff. She had asked to be cremated after her death and that her ashes be added to Metchnikoff's urn in the Pasteur Institute library. ("There's room there," she explained in a letter to her brother.) Probably because of the turmoil of the war, her request was not granted. She is buried in the Bagneux cemetery southwest of Paris.

As for the four hundred letters to her, after World War II ended, one of Olga's sisters-in-law, with the help of the Soviet embassy, managed to have them retrieved from the south of France and transferred to Moscow, where they are now kept in the Metchnikoff fund in the Archive of the Russian Academy of Sciences.

44

VANISHING

WHEN ALEXANDRA TOLSTAYA, LEO TOLSTOY'S daughter, visited Japan in the 1920s, she was surprised to hear Metchnikoff's name, altered by the Japanese accent, being shouted in the street. "Every morning I watched a boy in a spick-and-span white gown and cap deliver Metchnikoff's yogurt on a bicycle, yelling at the top of his lungs: 'Irya, Irya!'—that is, Ilya Metchnikoff (the *r* and the *l* sounding nearly the same)," she recalled in an autobiographical book. Metchnikoff's legacy became even more closely linked with the Japanese dairy industry when in 1935, inspired by his work, Kyoto microbiologist Minoru Shirota created Yakult, a fermented milk drink deliberately loaded with lactic acid bacteria.

Possibly as a result of the craze created by Metchnikoff, yogurt had turned into the most popular of all types of fermented milk. And if Metchnikoff's name remained a household word for a decade or two after his death, it was mainly in connection with yogurt—just as the *Scientific American* had predicted. Le Ferment had closed after Metchnikoff died and Émile Rémy had been mobilized, but other companies, some of them later to turn into dairy giants, continued to endorse their products with Metchnikoff's name.

In 1919 in Barcelona, a small business called Danone (and later Dannon in the United States) started selling sour milk in clay pots through pharmacies, mainly as a remedy for children with intestinal disorders. "We used tinned-copper vats to heat the milk and stirred it by hand with wooden paddles," Daniel Carasso, the son of Danone's founder, is quoted as recalling in the company's historical brochure. "Then we added the yogurt culture with a pipette, one jar at a time. We used a culture from a strain supplied by the Pasteur Institute." Ten years later in Paris, these pots made their way into dairy-food shops, carrying the sticker, "Danone yogurt. Prepared as recommended by Professor Metchnikoff of the Pasteur Institute and recommended by the medical profession."

Also in 1919, in the United States, John Harvey Kellogg argued in his book *Autointoxication* that Metchnikoff's early demise in no way discredited his theory of longevity. "Thousands of persons felt a keen sense of disappointment when the news of the great scientist's death was flashed across the Atlantic," Kellogg wrote. "Had the Metchnikoff philosophy failed? We answer, No; most emphatically, NO." In the chapter "Metchnikoff's Mistake," he declared that consuming milk-souring germs was not enough: Metchnikoff should have kept his intestines squeaky-clean by means of enemas, in which Kellogg himself religiously believed.

A few years later the advent of sulfa drugs, and then antibiotics, pushed sour milk aside as a remedy. The drugs were so effective at killing the microbes inhabiting the body, it seems that they also killed the interest in the gut flora. The new emphasis was on eliminating intestinal germs, not altering them. If Metchnikoff's ideas on intestinal flora and aging were remembered from time to time, it was mainly to be put down. "It would seem that in his views on the production of senile changes in the tissues caused, as he thought, by intoxication from harmful intestinal bacteria, his imaginative powers outran his judgment," *Nature* wrote in a rare article about Metchnikoff in the early 1940s.

As for immunology, not just the study of phagocytes but immunity research in general went into a decline for a while after World War I. Serum therapy hadn't lived up to its expectations of a panacea. Besides, in the words of science historian Arthur Silverstein, immunology had run out of easy successes. To the extent that studies on immune mechanisms did continue, they mainly focused on antibodies. Metchnikoff's scientific heritage faded in world research, as did he himself in the public mind.

Just the opposite happened in his homeland.

In Mikhail Bulgakov's novel *The Heart of a Dog*, written in the mid-1920s, Moscow surgeon Filip Filipovich Preobrazhensky, famous for working wonders in rejuvenation, comes home to discover that the mutt he had picked up in the street had chewed up his stuffed owl and smashed a glass-framed photograph on his desk. "Why did you break Professor Metchnikoff?" Preobrazhensky lashes out at the dog.

In the former Russian Empire, reminted as the Soviet Union, Metchnikoff's portrait was indeed de rigueur on the desks of many real-life scientists, as it was on the walls of many real-life institutions. In fact, an ironic, even absurd thing had happened. Almost immediately after the Bolshevik revolution, Metchnikoff had been inducted into its ideological hall of fame—a pantheon of historic personages turned into cult figures by Soviet propaganda. Had he lived a bit longer, he would have found himself among the revolution's victims, stripped of the income that had enabled him to engage in research. Besides, as a landowner and intellectual, complete with spectacles and a three-piece suit, he cut the quintessential figure of a *bourzhouy*, a "bourgeois," who in the eyes of the Bolsheviks deserved to be hung from lampposts. To top it all off, his Jewish roots, downplayed in Soviet biographies, were anathema to the anti-Semitic regime.

But Metchnikoff was dead, so he had no say in the matter.

Together with other prominent Russian scientists, generals, writers, and musicians, not to mention Lenin himself and other Bolshevik leaders, he was recruited into the erasing of Russia's tsarist past. In St. Petersburg, or Petrograd, soon to become Leningrad, the enormous Peter the Great Hospital was renamed after Metchnikoff. Various other institutions, professional societies, and dozens of avenues, streets, and lanes throughout the Soviet state were also given his name. Most ironic was probably the naming after him of the two Odessa institutions on which he had slammed the door before leaving Russia: the bacteriological station and the Novorossiya University, today the Odessa I. I. Metchnikoff National University.

For all the irony, it was almost inevitable that Metchnikoff would become one of his homeland's scientific heroes. Since dozens of Russian doctors and scientists had attended his courses at the Pasteur Institute or received training in his lab, an entire first generation of Soviet bacteriologists was made up of his grateful former students. Besides, he and Ivan Pavlov were—and still are, at the time of this writing—the only Russian scientists ever to have won a Nobel Prize in Physiology or Medicine.

In May 1945, the one hundredth anniversary of Metchnikoff's birth was celebrated in the Soviet Union on such a heroic note, it was as though the biologist had personally contributed to the recent victory over Nazi Germany. In the Kremlin, the Council of People's Commissars passed a decree to commemorate him in every imaginable way: prizes, scholarships, medals, plaques, statues, a biography, a film, and publication of his collected works. The health minister himself opened a three-day symposium in his honor in Moscow and eulogized him in *Pravda*, proclaiming, "Metchnikoff's works have brought glory to the Russian people, reflecting the richness and greatness of Russian culture, as well as Russia's enormous contribution to world science."

When the Soviet Union soon marched into the Cold War, Metchnikoff was yet again among the Russian luminaries called to the battleground to prove Soviet superiority over the West. That was why I later encountered him in my high school textbook of history, not biology. He was featured among the great historic figures of "Russian culture in the imperialist epoch." As I realized later, I was wrong in dismissing him together with the entire textbook, but I was right in being suspicious about the way he was being presented. He had been co-opted and taken out of context for ideological purposes, at a time when his legacy was out of fashion in world science.

This discrepancy is evident in the sixteen volumes of Metchnikoff's *Academic Collection of Works*, which began to appear in the 1950s in line with the commissars' decree. An editorial preface to his writings on immunity stated, "I. I. Metchnikoff's works serve as a constant reference for Soviet immunologists." That was not at all true for immunologists outside of the Soviet Union. In fact, on the occasion of the 125th anniversary of Metchnikoff's birth, the Russian-language journal

Mikrobiologia complained about "attempts by certain bourgeois scientists of Europe and America to hush up the immortal accomplishments of the Russian biologist," implying that the neglect of Metchnikoff was ideologically motivated. The journal even quoted Pushkin to the effect that "disrespect for ancestors is the first sign of barbarity and immorality."

In truth, Metchnikoff's oblivion had nothing to do with barbarity or morality. It was just that when immunology research picked up in the second half of the twentieth century, phagocytosis was largely neglected. Antibodies were still viewed as the be-all and end-all of immunity.

At the 1954 annual meeting of the American Pediatric Society, Boston doctors described a mysterious syndrome: five babies and young children had come down with so many infections they ultimately died. The physicians called it "an immunological paradox." The children got sick despite having high levels of antibodies in their blood. Several doctors in the audience said they had encountered similar cases. One San Francisco physician mentioned that the blood of some of these children almost lacked certain phagocytic cells. No one paid attention. It would take years for the "paradoxical" immune deficiency—now called chronic granulomatous disease, or CGD—to be linked to the malfunction, and sometimes also shortage, of those phagocytes.

Even when the role of cells in immunity eventually did gain prominence in immunology research in the 1960s and 1970s, the focus was not on Metchnikoff's phagocytes but on the little round cells called *lymphocytes*, the T and B cells, which were much easier to obtain and grow. In the words of one prominent researcher, all of immunology then boiled down to the question "T-B or not T-B?"

In 1980 a handful of researchers who did carry Metchnikoff's torch over the years convened at a conference, "A Century Since Metchnikoff: Phagocytosis—Past and Future," in the Sicilian town of Taormina, about an hour's drive from Messina. To emphasize the Metchnikoff connection, the book of the conference's proceedings, published two years later, stated the meeting's location as "the Province of Messina." But just like the conference's symbolic venue, the study of phagocytes was but an island in the sea of contemporary immunological investigations. The Messina meeting brought together a total of fifty participants from Europe, North America, and elsewhere, whereas at the time, the American Association of Immunologists alone had some three thousand members.

By then, immunologists had come to regard immunity as one complex entity called the immune system. But an entirely new subdivision within this system emerged, and continues to hold today. In this subdivision, the immune system is no longer viewed as divided into cells and body fluids. Rather, it is seen as consisting of two arms, innate and adaptive, each comprising both cells and various substances.

Metchnikoff is regarded in retrospect as the founding father of research into innate immunity, and Paul Ehrlich as the founder of studies into adaptive immunity. Innate immunity, the first to respond to a threat, is so called because it is present at birth and remains unchanged throughout a person's life; it is determined by the genes. Adaptive immunity takes longer to get organized because, as its name suggests, it adapts specifically to different threats. It improves during a person's life because it retains the memories of encounters with various attackers.

But repeating the outcome of the immunity war, this new subdivision initially made Metchnikoff look less important than Ehrlich. Innate immunity involves the most ancient immune mechanisms, including phagocytes—ones that we share with primordial cousins such as insects, worms, mollusks, and other invertebrates. It probably appeared along with multicellular organisms about a billion years ago. Adaptive immunity involves defenses that exist only in vertebrates, including antibodies and T and B cells, which means that it is "only" about 450 million years old. "Metchnikoff's" innate immunity, an ancient collection of mechanisms shared with simpler organisms, was at first seen as more primitive than "Ehrlich's" more recent adaptive one. Besides, adaptive immunity mechanisms are more targeted (*specific*, in technical terms) and endowed with a memory. They were viewed as more sophisticated and worthy of study.

When I talked to people who did work on "Metchnikoff's" immunity in the 1970s and 1980s, they all recalled the scientific community's predominant interest in adaptive defenses. "The antibodies and later the T cells clearly captivated more people, I'd say ninety-five percent of the people in terms of research," Ralph Steinman, who would soon receive a Nobel Prize for his immunology research, told me in an interview. Ira Mellman, another prominent veteran immunologist, concurred. "In the 1980s, the majority of smart people in immunology were working on T and B cells," he said.

Mellman, who himself had been among the smart people in the minority, said there were profound reasons why innate immunity remained outside the mainstream for so long:

> The biggest conceptual challenge was understanding the diversity of antibodies: how can your body react to a limitless number of foreign agents? That was an enormous philosophical problem, up there with understanding how DNA encodes inheritance. It drove the field for decades. Phagocytes were thought to be dumb and boring; there was no reason to care about them. All they did, it seemed, was to buy time until the elite troops of the T and B cells arrived. Besides, incorrectly, innate immunity was thought to be entirely separate from adaptive.

This situation prevailed even at the Pasteur Institute, which over the years had come to occupy a densely built-up campus in a bustling Parisian neighborhood. In its massive red-brick, six-floor Elie Metchnikoff Building, which according to the sign outside "is home to a large part of immunology research on campus," nobody worked on "Metchnikoff's" innate immunity. As for Metchnikoff himself, he suffered the indignity of being a founder of a relatively unimportant field. Of course, the glory of the Nobel Prize was his forever, but he remained an underdog in the rarified circles of Nobel laureates. If he wasn't entirely forgotten, he and his phagocytes were often snubbed.

But with such a profound vanishing act would come an even more striking comeback.

V

LEGACY

45

METCHNIKOFF'S LIFE

The austere steppe around Panassovka, renamed Mechnikovo, where Metchnikoff spent his childhood, hasn't changed much since he explored it as a boy more than 150 years ago. Carpeted by grass and wormwood, ever fragrant with herb-filled aromas, its stillness is rarely disturbed by anything but the gaggling of geese or the buzzing of flies. It's a landscape in which nothing limits the flight of fancy, a fitting backdrop for the imagination that guided Metchnikoff's research.

When I visited Mechnikovo in 2009, I found a quaint, semideserted village smelling of roasted sunflower seeds. Compact whitewashed houses silently lined its total of three streets. Many had single occupants, others were inhabited only in the summer. The village's young people had headed to the city to look for jobs. The remaining residents, most of them retired, were hard put to adjust to a market economy, leasing their lands in exchange for barley and wheat.

In his living room, where a portrait of Lenin still hung, former local secretary of the Communist Party Vasily Nikolayevich Stroyev spoke with nostalgia about the Soviet days when Mechnikovo had bustled with life as the cradle of a national icon. Now movies, concerts, and football games were all history. So was the Metchnikoff Elementary School and even the grocery store. Not much remained here of Metchnikoff other than his name. An exhibition of his photographs and documents, pinned to burlap-fitted frames, gathered dust in Mechnikovo's dilapidated, locked-up club. The brick remnants of the family's two-story house, picked apart by peasants decades ago, had been long overtaken by tall grass. The days of profligate state sponsorship for national symbols were clearly over.

On this and other research trips to Ukraine and Russia, I found that many young people who went to high school in the post-Soviet period had never even heard about Metchnikoff. Still, in official circles, Metchnikoff, just as he did after the Bolshevik revolution, again fills the need for heroes untainted by the deposed regime—this time, the Soviet regime. In Kharkov I admired his red-granite sculpture, reminiscent of the Soviet hero-worship style, erected in front of the Metchnikoff Institute of Microbiology and Immunology as recently as in 2005, the 160th anniversary of his birth. The Russian Academy of Natural Sciences has lately instituted a Nobel Laureate Ilya Ilyich Metchnikoff Honorary Medal "for practical contributions to the enhancement of national health." In 2011 Metchnikoff's name was awarded to the newly reorganized North-Western State Medical University in St. Petersburg.

His name lends prestige not only to academic institutions. In a sign of the times, the Metchnikoff Moscow Bacteriology Institute created by his former students in 1919, the one that became home to his museum (much of its collection has since been moved to other museums and archives), is now the Metchnikoff Biomed biotechnology company, manufacturing vaccines and probiotics. And it is not only the Russians who claim Metchnikoff as their own. Not so long ago, he was featured in a Ukrainian television series, *100 Great Ukrainians*, and because his mother was Jewish, he was included in a Russian-language dictionary, *Jewish Nobel Laureates*.

As for the global fermented milks industry, you'd think it would no longer need Metchnikoff. You'd be wrong. My Internet search has turned up "Metchnikoff's Life," a fermented milk drink made by South Korea's Hankuk Yakult that proudly displays Metchnikoff's patriarchal portrait on the cup. A recent Russian-language commercial for Danone yogurt featured a radiant Metchnikoff, played by a bearded actor. And since 2007 the Brussels-based International Dairy Federation, or IDF, has been awarding an IDF Elie Metchnikoff Prize to promote research "in the fields of microbiology, biotechnology, nutrition, and health with regard to fermented milks."

In parallel with discovering ways in which Metchnikoff was still remembered in his homeland and by the international dairy industry, I traced the fate of Lili.

After World War I she moved with her parents to Tunisia for a while, where she married Raoul Saada, a Jewish lawyer whose wealthy family owned a large carpet store in Tunis. At the request of Raoul's family, she converted to Judaism but never practiced it afterward. She later returned to France with her husband and their only son, Jacques, settling in Ville d'Avray, adjacent to Sèvres. She stayed in touch with the Pasteur Institute and maintained an emotional correspondence with Metchnikoff scholars in the Soviet Union. Together with Jacques, she regularly laid flowers at Metchnikoff's urn in the Pasteur Institute library on anniversaries of his birth and death, until her own death in 1982. Her red-granite tombstone in Ville d'Avray's flower-studded cemetery bears the inscription *Filleule adorée de Elie Metchnikoff*, "Adored goddaughter of Elie Metchnikoff." It's unknown whether the inscription had been her or Jacques's idea, but the Metchnikoff connection had clearly been central to her self-identity.

At the same time as I was trying to get access to Lili's Metchnikoff archive that Jacques Saada had placed in the bank on the Champs-Elysées, I also tried to learn as much as I could about Saada himself. After all, as Lili's son, he had been Metchnikoff's last living legacy. My initial source was the soft-spoken Dr. Patrice Rambert, whose name I found on Saada's death certificate. He had treated Lili as a physician and had taken care of Jacques in the last few miserable years of his life. Thanks to the leads Dr. Rambert provided, I interviewed Jacques's other acquaintances and friends and was ultimately able to retrace the heartbreaking story of Metchnikoff's purported grandson.

I found myself transported from the Pasteur Institute to Hollywood. After years of science writing, I was suddenly immersed in the semihallucinatory world of a fanatic cinema buff, rife with silver-screen characters and twists of plot.

Jacques Saada had been a lawyer who never held a regular job, a gallant gentleman who never had a full-fledged romantic relationship, a dreamer whose grandiose schemes were destined to flop. He was a wishful thinker like Metchnikoff, but with none of his purported grandfather's knack for converting his wishes into fruitful undertakings.

Born in Tunisia, Saada studied law at the University of Paris after moving with his parents to France. What truly moved him were music and art. Inspired by the family's Russian connection—that is, Metchnikoff—he represented in France the copyrights of heirs to various Russian composers and artists. But his biggest passion of all was the cinema. He dreamed of making his own movie.

Ebullient, eloquent, and brilliantly intellectual, the dapper, bespectacled Saada, tall and slightly overweight, had the habits of someone who came from money. But he had more of a flair for spending—he wined and dined friends with style—than for making money. Repeatedly expecting to make a fortune, he inevitably ended up empty-handed. After both his parents died in the 1980s, he started squandering his substantial inheritance. He chose a hero as grand as his own ambitions, Napoleon Bonaparte, and began making a film about him, buying expensive equipment and bringing over a camera crew from Russia but never completing the project.

Never married and having no children, Saada, as befits a cinema devotee, was madly in love with a Hollywood star. For him, each film featuring Sharon Stone, then at the height of her fame, was a festive occasion. He invited friends to see these films in a movie house with the largest screen possible and even met Stone in Paris, dining with her and her entourage in the legendary *Le Moulin de la Galette* on Montmartre. Contributing raving entries about her movies to the popular dictionary *Guide des films*, Saada extolled Stone as an untouchable goddess. With time, his behavior grew progressively eccentric. His Sharon Stone fetish became a joke among the regulars at Paris's cinematheque.

Disaster struck when in the late 1990s Saada ventured into making a documentary, *Sharon Stone, la Chevalière*, presenting Stone as a descendant of the Teutonic chevaliers. The enterprise ruined him. He incurred enormous debts, which were not covered even by the sale of his Ville d'Avray apartment. Saada no longer had a home. Mortified, he was reduced to borrowing money from relatives and friends.

But he remained upbeat. His next big project was to sell his Metchnikoff collection—Metchnikoff's letters, papers, and personal objects that Lili had preserved. He had the materials evaluated at Sotheby's with an eye to an auction, although he really hoped to sell the collection in Russia, where Metchnikoff was still a national hero. The authorities there, he believed, would pay millions to return to their country this valuable portion of national heritage. He placed the collection in the safe deposit boxes at the Crédit Lyonnais bank—not unexpectedly, he chose a branch in a fancy spot, on the Champs-Elysées—and started negotiating with "the Russians" through a former Russian diplomat living in Paris.

One day, coming downstairs from the studio he was renting in a hotel on a small, quiet square by the Sèvres-Ville d'Avray train station—the same station where Metchnikoff used to get off after taking the Saint-Lazare line back home from Paris—Saada broke a leg. The fracture had resulted from a tumor. The diagnosis: bone cancer. He made out a will, leaving his most precious belongings—the Metchnikoff collection and the rights to the Sharon Stone documentary—to a tall, blonde Latvian fashion journalist half his age, whom he had met on a trip to the Soviet Union and invited to Paris on a number of occasions.

The deal with "the Russians" had fallen through, in the manner of Saada's other pipe dreams. After he died at age sixty-eight in July 2003, the Metchnikoff collection remained in the safe deposit boxes at the Crédit Lyonnais. The boxes were opened in the fall of 2012—when I was allowed to make photocopies of documents in the collection for a few hours—and then closed again with new locks. When this book went to print, the Metchnikoff collection was still locked up in the bank vault.

46

METCHNIKOFF'S POLICEMEN

In January 2010 I attended a popular lecture, "War Movies: A Close Look at the Immune System's Struggle Against Infection and Cancer," at the Weizmann Institute of Science in Israel, where I work as a science writer. I was then still planning to write a book about the crying injustice done to Metchnikoff's memory, but at that time, as I began to realize at the lecture, history was already writing an entirely different ending to his story.

To my surprise and delight, the very first slide to light up the screen was a portrait of a shaggy-haired Metchnikoff. The lecturer, a young immunologist named Guy Shakhar, began a talk about the latest advances in his field with an homage to Metchnikoff's discovery of phagocytosis. The audience burst into laughter when Shakhar played a real-life video of a phagocyte that seemed to be stubbornly pursuing a germ to the sounds of chase-scene music, like a cartoon character on a mission. No mystical vitalism, of which Metchnikoff had once been accused, was involved. Rather, the phagocyte was drawn to the chemicals released by the germ. Metchnikoff could have spared himself years of anguish if only he could have shown such a video to his scientific opponents early on in his battles.

Examining current scientific literature, I discovered glorious recent tributes to Metchnikoff, marking the centennial of his Nobel Prize. In the journal *Nature Immunology*, I found a detailed report on the festive two-day symposium "Metchnikoff's Legacy in 2008," which was held at the Pasteur Institute in Paris. In other journals and on the Internet, I kept running into Metchnikoff's bespectacled image: on the cover of one of the issues of the *Journal of Leukocyte Biology*; in a reverent 2008 commemoration in the *European Journal of Immunology*, "Elie Metchnikoff: Father of Natural Immunity," by veteran macrophage researcher Siamon Gordon; in the first frames of an online iBiology.org seminar by Ira Mellman, "Cellular Basis of the Immune Response."

Definitive proof of a Metchnikoff revival in modern science came in 2011. When that year's Nobel Prize in Physiology or Medicine was announced, few people, apart from the laureates themselves, may have been as excited as I was. Two of the winners, Jules Hoffmann and Bruce Beutler, were recognized "for their discoveries concerning the activation of innate immunity"—that is, the activation of "Metchnikoff's" immunity. They had identified receptors called TLRs, which enable phagocytes and other cells to recognize such foreign invaders as bacteria, viruses, or fungi. This was the first time since Metchnikoff that the Nobel Prize was given for advances in innate immunity—a sure sign the field had finally moved to center stage in contemporary science. In the preceding century-long interval, all the ten prizes in immunology had been awarded for research on "Ehrlich's" adaptive immunity.

Hoffmann, who like Metchnikoff had conducted his prize-winning research on invertebrates (grasshoppers were among his favorites), later told me in an interview, "Metchnikoff made the first essential step in what's now called innate immunity research. Then came a long Siberian winter." He was talking about the decades during which this area of study had been largely neglected. It had ultimately been awakened around the turn of the twenty-first century, thanks to improved tools for handling cells and studying the molecular basis of immunity. The awakening was helped by the parallel emergence of new theoretical concepts and the foresight of numerous scientists; in particular, practically everyone I interviewed credited American immunologist Charles Janeway, who died in 2003, with pioneering the modern studies of innate immunity. By the mid-2000s, scientific journals were discussing a "rediscovery" of the innate component of the immune response. "This area lagged behind for so long, it now has to make up for that," Hoffmann explained.

One reliable way to determine whether a scientific field is "hot" is by the number of studies it generates. By this measure, in the twenty-first century, innate immunity research is in full swing. As many as some twenty-five thousand scientific papers have centered on TLRs alone in the roughly two decades since their discovery in the end of the twentieth century. Hoffmann believes this massive effort will help improve vaccinations and treat a slew of common diseases, particularly allergies and inflammatory and autoimmune disorders.

The study of various phagocytes has suddenly exploded too. A search for the topic "phagocyte" in the US National Library of Medicine's PubMed Central, an international database of biomedical research, turns up four studies published in 1930, thirty studies in 1960, and about four thousand studies in 2015. A search for "macrophage" yields three studies in 1930, thirty-four studies in 1960, and more than twelve thousand studies in 2015.

Metchnikoff would have been 166 years old when the Nobel Prize was awarded for research into innate immunity, no inconceivable age by his own standards. I tried to imagine him at the celebrations in Stockholm, how elated he would have been at the visual climax of Hoffmann's Nobel lecture—a gruesomely gorgeous electron micrograph of a dead fruit fly lacking TLRs. It had lost its battle against a fungal infection, so its upper body was veiled in hairy projections of fungi that looked like unkempt fur. Nearly 130 years earlier, in a friend's aquarium in Odessa, it was Metchnikoff's observations of exactly the same kind of battle—between an invertebrate (a water flea) and a fungus—that had given him strength throughout his immunity war. Indeed, in his Nobel lecture Hoffmann began the historical overview with Metchnikoff's work. When elaborating later on his own research, he described an antimicrobial protein fragment he had identified in fruit flies in 1995. He had named it Metchnikowin.

The third 2011 laureate in medicine, Ralph Steinman, was honored for his discovery of a previously unknown type of cell called *dendritic*, which in terms of Metchnikoff's immunity war could be classed as peacemakers. These cells provide a link between innate and adaptive immunity. They devour infectious organisms just as other phagocytic cells do, but in contrast to other phagocytes, their major role is not to destroy the invader but to alert additional troops, those of adaptive immunity. Thus, in contrast with the 1908 Nobel Prize, which recognized Metchnikoff's and Ehrlich's views of immunity as completely separate, the 2011 prize put a spotlight on communication within the immune system. "It's really silly—as Metchnikoff and Ehrlich tried to do when they defended their respective theories, to argue that innate immunity is more important than adaptive or the other way around. They are both important for different reasons," Steinman told me in an interview two years before he received the Nobel Prize. (He died two days before the prize announcement.)

Innate and adaptive immune mechanisms work together in a way Metchnikoff and Ehrlich never did. Innate immunity is the body's first line of defense. Springing into action immediately upon sensing danger, it lunges upon bacteria, fungi, or other offenders. For example, Shakhar, whose lecture first alerted me to the Metchnikoff revival, has shown that within half an hour of infecting the intestines, *Salmonella* germs are gobbled up in the gut lining by phagocyte-like dendritic cells, which belong to innate immunity but provide a crucial link to adaptive defenses. Video clips filmed under the microscope reveal these cells squeezing through the upper layer of the lining and sending their extensions, the dendrites for which they are named, to capture the bacteria. The cells then rush the germs to the nearby lymph node to recruit the tools of adaptive immunity, the T and B cells, which take a few days to get organized but ultimately mount a targeted defense: B cells produce antibodies exquisitely tailored to fight the *Salmonella* germs. Thus, Metchnikoff's insistence on the phagocytes' involvement in producing antibodies proved true in a certain sense in some cases, even if the production itself turned out to be the work of other cells.

In his other claims about phagocytes, he was often amazingly prescient. He has been fully vindicated, for instance, in his belief that macrophages play a central role not only in immunity but also in the maintenance of body tissues. Giving him ample credit for this claim, one review on macrophages in the journal *Trends in Molecular Medicine* has been aptly titled "Metchnikoff's Policemen."

Strategically placed throughout the body tissues, macrophages, in addition to gobbling up infectious organisms, swallow dead cells and debris. They take part in the continuous turnover of bone, help remodel the connections between neurons in the brain, and facilitate tissue regeneration in the liver. Some scientists believe macrophages may well be the most abundant single cell type in the body. Every gram of liver tissue, for instance, contains millions of macrophages, which make up about one-fifth of all liver cells. And these "big eaters" don't just eat; they release a host of substances that regulate blood pressure, control fat metabolism, and promote tissue repair.

But Metchnikoff's policemen sometimes become dangerous. They don't always know when to stop, something like Metchnikoff himself, so that a beneficial biological process leads to damage. "Macrophages are involved in almost every disease," states a review in the journal *Nature.* These cells can release substances that promote insulin resistance, fat accumulation, and inflammation thought to contribute to such neurodegenerative diseases as Alzheimer's. The plaques in atherosclerosis are mainly made up by macrophages stuffed with cholesterol. Alas, Metchnikoff's policemen may have hastened his own death of heart disease. The substances they produce can cause scarring, or fibrosis, in the heart muscle. He himself traced the origin of his heart disease to his self-inoculation with relapsing fever in 1881. His autopsy indeed revealed that his heart had pathological changes of a long-standing nature.

On a brighter note, in line with Metchnikoff's utopian dreams, the pervasive role macrophages play in disease has opened up an entirely new and unexpected direction in the search for future therapies: targeting the macrophages. Perhaps most surprising is an innovative line of research aimed at treating cancer.

Initially, when scientists found large numbers of macrophages in cancerous tumors, they assumed these cells were attempting to fight the malignancy. In fact, generally, macrophages eliminate newly formed cancerous cells throughout the body. But as it turned out, some tumors manage to not only evade this defense but also "corrupt" the macrophages, tricking them into helping the tumor grow and spread. Experimental drugs aimed at blocking these "corrupted policemen," as one scientist calls them, are already being tested in patients.

47

ULTIMATE CLOSURE

When I looked into ways in which phagocytosis research is already transforming people's lives, the most moving experience was my interview with a thirty-two-year-old woman I'll call Dina. Welcoming me into the cozy apartment she shared with her husband and three-year-old daughter in a southern suburb of Tel Aviv, she appeared confident and healthy, with shiny raven-black hair, chiseled features, and slightly slanted olive eyes. Only a faint scar on her neck, marring her otherwise porcelain complexion, gave away past troubles.

"I never imagined I'd get married. I didn't believe anybody could love me that way," she told me. "And I was certain I'd never have kids. I just couldn't face the responsibility."

As a child, growing up in northern Caucasus, Russia, Dina repeatedly came down with pneumonia and swollen lymph nodes. Because of the chronic inflammation in her lungs, doctors "diagnosed" her with tuberculosis, even though they never found its germs in her body. Throughout her childhood, she spent winters in hospitals and sanatoriums, swallowing pills by the handful and getting shots against tuberculosis. Because of the constant illnesses and moving about, she made no friends and grew so withdrawn that in third grade, teachers thought she was intellectually disabled.

Things did not improve much after she'd moved to Israel with her mother at age eleven. She had an operation on her neck to drain an abscess, kept coming down with boils all over her body, and was once again treated for tuberculosis. Other children avoided her because they thought she was contagious, teasing her when she wore a scarf in the summer to hide her scars. "When I was little, I didn't think I'd have a normal life. I had terribly low self-esteem," she recalled.

At twenty-two, as a result of a severe lung infection, she had an entire lobe of one of her lungs removed. "I had a huge scar on my side. I thought they'd always be cutting me until I'd look like Frankenstein." After three months in the hospital, she had lost twenty-four pounds and was so weak she couldn't walk up the stairs. She sank into a severe depression.

Just then, one of the doctors at the hospital decided to send her blood for analysis to a lab specializing in immunological diagnosis. Dina was found to suffer from the same immune deficiency that had been termed "an immunological paradox" at the 1954 meeting of the American Pediatric Society, now called chronic granulomatous disease, or CGD, and known to be caused by a malfunction of phagocytes. She had inherited a genetic mutation responsible for this malfunction from both her parents, who were second cousins. Neither of them was sick, because each carried only one copy of the mutated gene, but Dina, having the two copies, ended up with impaired immunity.

The full impact of having a diagnosis did not immediately register with her. "The doctors said it was a good thing to finally know what's wrong with me, but I thought to myself, *So what?*" she said. Then she realized that a year had gone by, then another, and she was no longer getting sick. As she has a relatively mild form of CGD, preventive antibiotics could control the disease, and any signs of emerging illnesses could be instantly eliminated with drugs suited to treating the infections characteristic of CGD. Her self-confidence grew gradually.

At twenty-seven, she got married. Her daughter was born two years later. At the time of our meeting, she was expecting her second child. She had originally earned a degree in biology, motivated by the desire to understand her own disease, but later switched to an area in which she'd always wanted to work: interior design. "All I've been through has made me a very happy person because I value the things I thought I'd never have—a husband, a family, a good job," she said. "I can't even compare my life before and after the diagnosis; it's a different world, physically and mentally. Now I know that nothing is wrong with *all* of me, that only something concrete is the matter, and I know how to deal with it."

In Dina and others with CGD—a rare condition affecting about 1 in 250,000 newborns in Europe and the United States—that "something concrete" is a genetic mutation that impairs the function of phagocytes. The phagocytes have difficulty killing the invaders and then accumulate in infected tissue, along with other cells, creating the small nodules, or granulomas, that give the disease its name.

In the mid-1950s, shortly after the American Pediatric Society's meeting, the then-mysterious immune deficiency was defined as a syndrome called "fatal granulomatous disease of childhood." The children who suffered from it, plagued by endless sores, abscesses, and infections, died in the first few years of life. With improved treatments, the word *fatal* was dropped from the name, but it took another ten years for CGD to be confirmed as a disease of defective phagocytes.

Many of the symptoms in CGD arise from the impaired activity of the phagocytes called *neutrophils*, the most abundant type of white blood cell. They belong to the category of cells Metchnikoff called *microphages*, or "small eaters," a term no longer in use. Neutrophils are commonly used in diagnosis because they are easy to obtain from the patient's blood. Whereas Metchnikoff's larger "policemen," the macrophages, are long-lived phagocytes residing mainly in tissues, the much smaller neutrophils are short-lived, usually lasting less than a day, their supply continuously renewed by the bone marrow. When alerted by danger, they leave the bloodstream, migrating to the site of injury or infection. These "minor policemen" pounce on bacteria or fungi, engulfing and digesting them. Pus is a collection of dead neutrophils that sometimes accumulate at the site of the battle.

In the Meir Medical Center's Laboratory for Leukocyte Functions, Israel's center for CGD diagnosis and research where Dina's immune deficiency was identified, neutrophils are put through the mill to see whether they work properly, and if not, to determine exactly why not. "Metchnikoff talked about phagocytosis as a general process, but today we systematically check every step, to know where exactly to put the emphasis in therapy," said Professor Baruch Wolach, head of the laboratory, who has been investigating CGD for more than three decades.

As the first step, the neutrophil must recognize the invading germ and move in its direction, attracted to the chemicals the germ releases. Next, the cell has to stick to the blood vessel lining, squeeze through the vessel wall, and proceed to the focus of the infection, be it in the lung, the ear, the liver, or elsewhere in the body. Opsonins—the molecules that Almroth Wright discovered and George Bernard Shaw lampooned—facilitate this journey by marking the germ, amplifying the biochemical signal that attracts the neutrophil. Finally, the cell must engulf and kill the germ.

"I remember sitting in a large room with a confocal microscope, watching the movement of the neutrophil, how it captures the germ and the germ dies," Wolach said, describing the first time he observed phagocytosis through an advanced microscope. "It was exactly how Metchnikoff had described it, and he had none of the modern tools! I had tears in my eyes seeing that Metchnikoff's vision, that everything I'd learned and believed, was real."

Things can go wrong at each step of phagocytosis, but most common are defects in the killing mechanisms, and that is precisely the definition of CGD: phagocytes swallow the germs but are unable to destroy them. Honed by billions of years of evolution, these mechanisms are so complex that scientists are still working to unravel them fully. When the phagocyte receives a call to fight the germ, proteins within the cell assemble into a molecular complex that manufactures a series of enzymes, which, in turn, produce a major killing weapon—hydrogen peroxide, known from day-to-day life as a type of bleach or disinfectant. Other killing mechanisms are at play too. Because of their destructive power, the killing takes place inside an impervious pocket within the cell, to prevent damage to cellular structures.

A single mutation in any of the genes orchestrating this process can disrupt the killing. The severity of the resulting immune deficiency varies widely, depending on the particular genetic defect. Dina, for instance, was found to have a mutation in the gene encoding the enzyme that makes hydrogen peroxide. In a laboratory test, her neutrophils were able to destroy only about one-third of the germs they swallowed.

CGD makes it abundantly clear just how crucial phagocytes are to human health and even survival. And CGD is only one of many disorders in which various phagocyte defects endanger the patient's life. For example, in one such disorder, leukocyte adhesion deficiency, neutrophils fail to fight infection because they don't adhere to certain molecules, which prevents them from effectively "climbing" out of blood vessels and reaching infected tissues.

The crucial role of the phagocytes is no less evident from an immune deficiency of a different sort, in which the numbers of Metchnikoff's "minor policemen" are too small—a condition called *neutropenia.* Every drop of blood of a healthy person contains about seventy-five thousand neutrophils, or, in laboratory units, fifteen hundred cells per microliter. Any level below that is classed as neutropenia; it becomes life-threatening when the count plummets below five hundred. As in CGD, this can result from genetic defects, but more commonly, neutropenia occurs because the normally produced neutrophils are being destroyed, for example, as a side effect of cancer chemotherapy, which kills all rapidly growing cells. People with severe neutropenia are so vulnerable to infection, they are hospitalized in isolated chambers with strict sterile conditions.

In terms of the number of saved lives, neutropenia serves as probably the most tangible vindication to date for Metchnikoff's belief in his phagocytes. Current treatment of neutropenia—and no less important, its prevention—would be unthinkable if he hadn't established the role of phagocytes in immunity. Thanks to this understanding, various precautions now allow millions of cancer patients to receive lifesaving chemotherapy without falling victim to neutropenia. When necessary, patients can take drugs to boost the neutrophil count temporarily.

In CGD, in contrast, the underlying defect is there for life. In cases more severe than Dina's, when the phagocytes can hardly kill any germs at all, the only curative approach is bone marrow transplantation, itself a risky procedure that requires a matched donor. The patient's bone marrow is first destroyed, then seeded with stem cells that can give rise to properly functioning blood cells. An experimental alternative to bone marrow transplants is gene therapy, in which viruses are used to deliver to the bone marrow normal copies of the defective gene to enable phagocytes to do their job properly.

And this brings me to the part of the story that might have given Metchnikoff a bit of well-deserved glee. At the time of this writing, researchers conducting one of the leading gene therapy trials for CGD are working in Paul Ehrlich's former institute, Georg Speyer Haus, on Paul Ehrlich Strasse in Frankfurt, Germany. What better symbol for the cooperation between two arms of the immune system, innate and adaptive—and for the ultimate closure of the immunity war—than a "Metchnikoff" trial performed on an "Ehrlich" street?

48

LIVING TO 150

SEARCHING FOR CURRENT OPINIONS ON Metchnikoff's belief in people living to 150, I discovered that in 2001, two of America's foremost experts on aging had made a much-publicized wager precisely on this question: whether a baby who had already been born could live to the unprecedented age of 150. Biogerontologist Steve Austad, chair of biology at the University of Alabama, waged $150 to bet yes. Biodemographer S. Jay Olshansky, of the University of Illinois at Chicago, waged the same amount to bet no one was going to live that long. They invested the $300 in a trust fund. In the year 2150, under the terms of the wager, the offspring of the winning scientist will collect the fund, which by then they expect will have grown to half a billion dollars. Austad's descendants will get the money if at least one person on the planet, of sound body and mind, is found to have celebrated his or her 150th birthday. Olshansky's descendants will collect it if no such person can be found.

The wager goes to the core of the question Metchnikoff asked in *The Prolongation of Life*: "What is the maximum age a human being can reach?" Scientists have since come to the conclusion that there is no biological clock or built-in genetic program that sets a fixed limit on life span. Yet different species have characteristic average life spans, which does suggest the length of life is dictated by some biological mechanism.

As an ardent evolutionist, Metchnikoff would have appreciated that this mechanism is thought to be natural selection, the driving force of evolution. Theories proposed in the second half of the twentieth century, and still widely accepted in the twenty-first, hold that the life span of a species crucially depends on reproduction. Through natural selection, the species' biological makeup evolves so that its members live long enough to reproduce and rear their young to maturity, when they themselves can reproduce. Whatever happens at a later age is less important for the survival of the species, so traits that could potentially benefit older individuals are not "selected" from one generation to another. Worse still, selection no longer weeds out damaging traits, leading to the deterioration known as aging. We die of the resulting accumulation of damage.

Natural selection has shaped the life spans of different creatures to vary enormously. Mayflies typically live several hours; house mice, two to three years; bowhead whales, two hundred years. A specimen of an ocean quahog claims the longevity record for animals—this edible clam was found to be 405 years old. If the limits of life have proved so malleable in the course of evolution, it is logical to assume they could conceivably be stretched on demand. So what are the prospects of life extension for humans?

When I contacted Austad and Olshansky, some fifteen years after they had made their wager, both admiringly spoke of Metchnikoff as one of the earliest modern thinkers in the field of aging—and both told me they hadn't budged from their respective positions. "We do agree on one thing: nobody is going to live to 150 years unless we figure out how to directly affect the underlying rate of aging," Austad said.

It is in the underlying mechanisms of aging that Metchnikoff's major areas of research converge in unexpected ways. Metchnikoff would have been gratified—or horrified—to learn that in recent studies of aging, his macrophage policemen are yet again caught in the limelight. As people get older, these cells, along with other cell types in the brain, muscles, and other tissues, secrete abnormally high levels of inflammation-promoting substances—a process that has been dubbed "inflamm-aging." "We now know that aging is a state of gradually increasing inflammation," Austad said. "By toning down this rise in inflammation, we should be able to slow many aspects of aging, though probably not everything."

The current view, indeed, is that a single "cure" for aging is unlikely to be found. Rather, there might be a series of "cures" for the multiple failings that occur in aging. I've put "cure" in quotation marks because both Austad and Olshansky rejected Metchnikoff's definition of aging as a disease, explaining that scientists now view it as being entirely natural. Yet both conceded that in practical terms, such definitions are irrelevant. "Just because something is a natural process doesn't mean it can't be altered," Austad said. "Infections are natural processes too, and we can alter them by antibiotics."

"Curing" aging by altering the composition of the gut flora, as Metchnikoff had suggested, would surely be one attractive strategy. Tens of trillions of microbes belonging to thousands of species inhabit the human digestive tract. Together they weigh roughly as much as the brain, and their combined genes are estimated to outnumber those in the human genome by more than a hundred times. Modern studies into gut microbes have taken off with a vengeance in the twenty-first century, probably thanks in large part to the advent of fast and low-cost gene sequencing technologies. These have made it possible to launch several mega-undertakings, such as the Human Microbiome Project in the United States, aimed at sequencing the genes of all the human gut microbes, collectively known as the *microbiome*.

In this surge of studies, gut microbes have been linked to multiple aspects of health, including diseases of old age, but until very recently they were not connected to the aging process itself. Then, as in immunology, Metchnikoff was vindicated in his beliefs in this area. It turned out that the link between intestinal microbes and aging is not a "futile fantasy" after all.

The recognition came about in a truly Metchnikoffian way: through studies of invertebrates—worms and flies. Most of these studies were conducted in the second decade of this century, and it's still unknown to what extent their results apply to mammals—but they generated so much interest in the topic that a number of scientific meetings have recently been devoted to gut microbes and aging. In 2014, for instance, the prestigious biannual Cold Spring Harbor Laboratory's conference on the molecular genetics of aging for the first time included a session on the microbiome.

Even though the direct connection between the microbiome and aging remains to be clarified, Metchnikoff's ideas on this topic have already proved prescient in indirect ways: in modern discussions of large-scale measures for preventing age-related diseases. Experts are suggesting, for example, that the faltering immune systems of the elderly, a major source of ill health, may be boosted by manipulating their gut microbes, which, in turn, can probably be altered by diet. "The intestinal microbiome has become one of the hottest topics in the field of public health, and Metchnikoff speculated on that a hundred years ago. He was a hundred years ahead of his time," Olshansky told me.

If there is one area of gut flora research in which Metchnikoff is receiving full recognition, it is the study of *probiotics*, a term that applies to beneficial microbes and means "for life," as an antonym of *antibiotics.* So many probiotic products flood the market that it's not always easy to tell whether they promote health or hype, but since the end of the twentieth century, studies into probiotics have become part of the scientific mainstream. As an essay in the *Lancet* pointed out in 2012, two decades ago "attempts at 'probiotics' seemed a marginal practice, at best, consigned chiefly to unorthodox medicine," whereas currently they represent "both a serious field of medical research and a multibillion dollar global industry."

The *Lancet*'s essay is tellingly titled "The Art of Medicine: Metchnikoff and the Microbiome." I found numerous other current publications in which Metchnikoff's name is similarly placed next to a modern scientific term, apparently conveying a sense of wonder at the surprisingly modern studies he conducted a hundred years ago. One review, for example, was titled "Probiotics: 100 Years (1907–2007) After Elie Metchnikoff's Observation." In my interviews, scientists referred to Metchnikoff as "a mythical character in the field of probiotics" or "the grandfather of modern probiotics research."

As for aging, scientists are still trying to figure out how to slow it down, but human life spans have already zoomed up, exceeding expert predictions. In 2002 a paper in *Science* announced what it called "an astonishing fact": since the 1840s, life expectancy had risen at a steady pace of six hours per day, or two-and-a-half years per decade. The increase, the paper stated, had been "so extraordinarily linear that it may be the most remarkable regularity of mass endeavor ever observed." The trend held in the ten years that followed.

This is news in terms of human history. As noted in one scientific report, the bulk of the reduction in mortality had occurred since 1900, which means that it has been experienced by only about five of the thousands of human generations that have ever lived. During this period, the gain in life expectancy in western Europe and North America has been remarkable: about thirty years. In the United States, for example, between 1900 and 2010, it increased from forty-six to seventy-six for men, and from forty-eight to eighty-one for women. What's also new is that in the past, the average life spans soared mainly thanks to reduction in child mortality, but since the 1950s, the greatest gains in life expectancy have been among older people.

If mortality continues to decline at the same pace, most children born since the year 2000 will celebrate their one hundredth birthdays. That's a very big "if," and a topic of heated scientific debates. It would require such a major shift in human behavior toward healthier lifestyles that it is not too likely to happen but is not impossible.

Just as hotly debated is the upper limit of life span, as is amply illustrated by the Austad-Olshansky wager. Austad thinks the biological limits to the duration of life have been greatly overestimated. "We've been able to make mice live 75 percent longer than their average life span," he says. "To me that suggests the biological limits, if they are there, are much further out than we believe them to be. We have dozens and dozens of ways by which we can increase longevity in laboratory animals. I'm convinced at least some of these ways will turn out to be relevant to humans. I don't think we'll ever have a human average life span of 150, but having an occasional person live that long, say one hundred years from now, wouldn't surprise me at all."

Olshansky, on the other hand, believes the human body's biological design is such that it can't last 150 years. For one, some of the body's crucial components—muscle fibers and neurons—don't replicate. He strongly advocates trying to slow aging. He considers this the best form of disease prevention because aging is a contributing factor to many common diseases. But he thinks it's impossible to know how many years such efforts will add to human lives. "If you discover an intervention to modulate aging and believe it will help you live fifty years longer, you'd have to wait fifty years to know if it does," he says. Future gains in life span, he argues, are likely to grow progressively smaller.

My own bet is that Metchnikoff would have sided with the more upbeat view. "It is difficult to imagine that in the more or less distant future, science will not solve the issue of prolonging human life to the desired limit," he wrote in the preface to *Forty Years.*

Considering how imaginative his ideas always were, it is impressive to see how often they were shown to be scientifically solid. Metchnikoff "tried to force his beliefs down nature's throat," Paul de Kruif wrote in *Microbe Hunters.* "Strange to say, sometimes he was right, importantly right." Perhaps one day, in his faith in the prolongation of life, he will be proved right yet again.

Today his ultimate prize is probably the sheer momentum modern studies of aging have gained. The agony he felt on his deathbed, that his premature death was going to discredit his teachings, would have been eased had he known that gerontology, the field he invented and named, had turned into a thriving scientific endeavor. The thousands of studies conducted annually in this area may yet provide entirely new perspectives on his own research.

Elie Metchnikoff has been dead for a century, but his future may still lie ahead of him.

ACKNOWLEDGMENTS

When this book was still an idea, I turned for advice to the writer and author Rinna Samuel, legendary for her absolute editorial pitch. "If you want to write a book about Metchnikoff, you'll have to marry him," Rinna told me, referring to the level of intimacy with my subject that the project required. I felt honored that she'd agreed to help and appreciative that for several years, she had the patience to dissect one draft after another. She tried to teach me her own credo of the craft: that adjectives are to be used as sparingly as gold, that structure is everything, that less is more. Long after turning ninety, she occasionally critiqued this or that turn of phrase as "old-fashioned." My greatest regret is that Rinna passed away a month before her ninety-third birthday, and I didn't get a chance to show her the final manuscript. Her thoughts and comments are woven into the text. I could not have written this book without her crucial guidance.

Arthur Silverstein's *A History of Immunology* provided early inspiration for my research into Metchnikoff's "immunity war." As a historian of medicine well familiar with Metchnikoff's writings, Arthur Silverstein became an invaluable sounding board. For nearly ten years, he had the goodwill to answer endless e-mails, clarify medical concepts, help with translations from German and Latin, rescue me from scientific errors, and ultimately comment on the entire manuscript in rough form. Of course, I am solely responsible for the accuracy of the end result.

I am also greatly indebted to numerous other scientists and historians of science. Leslie Brent supplied the much-needed early encouragement. Alfred Tauber, whose writings about Metchnikoff were a source of continuous reference for me, placed his own collection of Metchnikoff's works at my disposal. Semyon Efimovich Reznik's Russian-language biography of Metchnikoff was a great resource; Semyon

Efimovich himself shared with me his insights into the history of Russian science, offering a fascinating perspective on *his* Metchnikoff versus *mine*. Oren Harman, of Bar-Ilan University's Science, Technology and Society Program, let me in on his delightful intellectual adventures in the history of science.

Many others answered questions by phone or e-mail, shared with me their own work, engaged in enlightening discussions, or spent hours explaining to me scientific concepts. They included Steven Austad, Paul Brey, Harald Bruessow, Jean-Marc Cavaillon, Kaare Christensen, Eran Elinav, Vadim Fraifeld, Vera Gorbunova, Siamon Gordon, Manuel Grez, Dan Heller, Jules Hoffmann, Steffen Jung, Mark Kipnis, Natalia Nikolaevna Kolotilova, Franz Luttenberger, Will Mair, Alberto Mantovani, Ira Mellman, Simin Meydani, Carol Moberg, Roald Nezlin, Erling Norrby, S. Jay Olshansky, Edgar Pick, Oleg Yaroslavovich Pilipchuk, Scott Podolsky, Aharon Rabinkov, Michel Rabinovitch, Gregor Reid, Ivan Roitt, Olivier Schwartz, Michael Sela, Guy Shakhar, Adi Shani, Elias Shezen, Ralph Steinman, Adrian Thrasher, Desmond Tobin, Anthony S. Travis, Tatiana Ivanovna Ulyankina, David Wallach, Gerald Weissmann, Baruch Wolach, and Elie Wollman.

In addition, Rachel Goldman and Morris Karnovsky shared their memories of the 1980 phagocyte meeting in the province of Messina; Silvie Melchior-Bonnet talked to me about her grandfather, Constantin Levaditi, and Raymonde deVroey about *her* grandfather, Jules Bordet; Ronit Gavrieli gave me a tour of the Laboratory for Leukocyte Functions at the Meir Medical Center; and Zeev Paikovsky, chief scientist of Tnuva, elaborated on the universal appeal of yogurt.

I would have never received access to the collection of Metchnikoff's papers locked up in the Crédit Lyonnais if Parisian lawyer Joseph Haddad hadn't managed to work his magic within the French legal system. He generously gave of his time to bring this move to a successful fruition. Thanks to his vast expertise in French law, not to mention his outstanding negotiation skills, he succeeded in obtaining permission to open the safe deposit boxes and arranged for the copying of the material in the smoothest possible manner. He was able to undertake his legal move thanks to the agreement of Erica Smelkova, Jacques Saada's heir. I am enormously grateful to her for her trust and cooperation. I'd also like to thank Maître Haddad's assistant, Muriel Garcia, for her efficient help.

The involvement of the Pasteur Institute was essential for obtaining the permission to copy the documents in the bank. I'm grateful to Henri Buc and Geneviève Milon, of the Comité Scientifique des Archives, for their crucial help and support, and to then director of the Pasteur Institute Alice Dautry for granting me exclusive rights to use the copies until the publication of this book. Copying this material within the few short hours at my disposal would have been unthinkable without the professional help of Daniel Demellier, who later provided me with ideal conditions for work in the institute's historic archives. Thanks also to the staff of the institute's library, Médiathèque, and Pasteur Museum, particularly Sandra Legout, a wizard of online resources, as well as Catherine Cecilio, Johann Chevillard, and Michael Davy.

"It's going to be a meeting you'll never forget," Annick Perrot, then director of the Pasteur Museum, wrote to me in 2007, after arranging for me to meet with Elie Wollman. I'm forever grateful to her for organizing that interview, as well as for answering numerous queries, locating historic photographs, and commenting on parts of my manuscript.

I owe special gratitude to Patrice Rambert, who helped me piece together Jacques Saada's story and showed me around Sèvres. Also in Sèvres, I was exceptionally grateful to Pascale Solere and Bruno Costa-Marini, who supplied me with one of the most moving experiences in my entire research by welcoming me into Metchnikoff's former house. For sharing with me their memories of Jacques Saada, I thank Alexandre Aristov, Agnes Pignol, Chantal and Gilles du Plessis, and Jean Tulard.

In my travels for this book, many people helped me make the most of each trip. Family and friends made sure my research trips to Russia were not only useful but enjoyable. My special thanks go to Alexander Borisovich Vikhanski and Tatiana Leonidovna Sharova for providing me with a warm home in Moscow, and to Valeria Dmitrievna Ugleva for her help with research in St. Petersburg. I'm grateful to Elena Efimovna Dinerstein for discussions of Russian history, and to her father, Efim Abramovich Dinerstein, for making helpful additions to the bibliography. In Italy, Christiane Groeben was a terrific host at the Naples

Zoological Station, which Metchnikoff loved to visit. I am also grateful to all those who helped me trace Metchnikoff's path in Mechnikovo, Kharkov, and Odessa, including the staff of the Zoological Museum at the Odessa I. I. Metchnikoff University, which has specimens dating back to Metchnikoff's time.

A great many people helped me conduct research in archives, libraries, institutions, and Internet communities in France, Israel, Poland, Russia, Ukraine, the United Kingdom, and the United States. In exploring the vast Metchnikoff fund in the Archive of the Russian Academy of Sciences in Moscow, I drew on the expertise of Irina Georgievna Tarakanova and enjoyed the welcoming atmosphere she creates for researchers in the reading room. Clarifying the ancestry of Metchnikoff's maternal grandmother, Catharine Michelson, was particularly challenging due to the paucity of records from the early nineteenth century, especially concerning women. I was able to track it down only thanks to clues supplied by Venedikt Grigoryevich Bem and to the enormous experience of St. Petersburg archivist Elena Evgenyevna Kniaseva.

Others who have been extremely helpful in responding to my queries or facilitating my research included Danielle Duizabo, La Colonie de Condé-sur-Vesgre; John Emrich, the American Association of Immunologists; Tatiana Gladkova, Bibliothèque Russe Ivan Tourguéniev, Paris; Lee R. Hiltzik, Rockefeller Archive Center; Merv Hobden, the "Microscopy as a Hobby or Profession" group at yahoo.com; Weizmann Institute librarians Anna Ilionski and Marina Sandler; Mykola Korinnyj, Derzhavnii Arkhiv Kievskoii Oblasti, Kiev; Hava Nowersztern and Maya Ulanovsky, Edelstein Collection, Jewish National and University Library, Jerusalem; Arlene Shaner, New York Academy of Medicine Library; Merav Segal, the Weizmann Archives; Rochelle Rubinstein, the Central Zionist Archives (Alexander Marmorek papers); and bibliography advisers at the *Leninka*, as Moscow's Russian State Library is still informally known.

Throughout my work, I could not stop marveling at amazing online resources, available free of charge, which saved me months, if not years, of research and travel. First and foremost, I'd like to express my appreciation for gallica.bn.fr, which provides searchable copies of French historic newspapers. Other valuable free resources included archive.

org, gutenberg.org, hathitrust.org, worldcat.org, and the Nomination Database at nobelprize.org.

During the Skidmore Summer Writers Institute, I benefited from having parts of the manuscript workshopped in the nonfiction class led by Honor Moore. I'm grateful to my fellow classmates for offering comments and providing supplementary material. Moreover, one of the greatest benefits of attending the Skidmore Institute was the opportunity of having a one-on-one tutorial with Lorrie Goldensohn, who gave the manuscript an immensely thoughtful and helpful reading.

Many friends and colleagues were generous with their time and help. Rachel Heller held my hand while I was figuring out how to get the book published. Tamar Vital commented on the proposal and June Leavitt on parts of the manuscript. Howard Smukler offered creative solutions for getting access to hard-to-find historic documents. Nurit Hermon identified plants in photographs of Metchnikoff's former garden. Eda Goldstein, Navah Haber-Schaim, Riza Jungreis, Anna Kiselev, and Oren Zehavi critiqued the entire book and made no end of helpful comments.

During the latest stages of my work, I had the pleasure of receiving guidance from Nancy Rawlinson, an uncommonly insightful and sensitive writing coach, who made valuable reading suggestions and offered sharp editorial advice.

My wonderful agent, Jessica Papin, at Dystel and Goderich Literary Management, helped me bring this book to print with great competence, enthusiasm, and a sense of fun and was always there for me long after I signed the contract.

At the Chicago Review Press, I'm grateful to Jerome Pohlen for his editorial acumen and for taking on a quirky project and to Michelle Williams and the rest of the staff for making the publishing process a smooth and pleasant experience.

Going further back in time, my gratitude goes to two remarkable teachers, Inna Aronovna Cohen and the late Anna Pavlovna Maslova, who many years ago managed to teach me English behind the Iron Curtain.

NOTES

Olga Metchnikoff's *Zhizn' Il'i Il'icha Mechnikova* (*Life of Elie Metchnikoff*) was often my source of information even when not directly cited. I used the Russian edition since that was the language of Olga's original manuscript. It was ready for publication in 1917, but because of the October revolution, the French version appeared first, in 1920, the English in 1921, and the Russian only in 1926.

When quoting Metchnikoff's books, I used the Russian or French versions, since translations into English, a language he knew less well, contain errors. For example, the English translation of *Immunity in Infective Diseases* states that he lectured at the 1890 Congress in Berlin, which he did not attend. All the translations from Russian, French, and German are mine; some of the German sources are cited from publications in Russian or English, as indicated.

In these notes, I use the Library of Congress transliteration system for Russian, but throughout the book, I provide phonetic spellings and spell well-known Russian names in their more familiar form—for example, Ilya rather than Il'ia and Tolstoy rather than Tolstoi.

Descriptions of weather in Paris are from *Le Figaro*, which in the late nineteenth and early twentieth century published daily reports on yesterday's weather.

FREQUENTLY CITED NAMES AND SOURCES

EM: Elie Metchnikoff

OM: Olga Metchnikoff

***ASS*:** I. I. Mechnikov. *Akademicheskoe sobranie sochinenii,* 16 volumes. Moscow: Medgiz, 1950–1964.

***Annales*:** *Annales de l'Institut Pasteur.*

***Letters 1*:** I. I. Mechnikov. *Pis'ma k Ol'ge Mechnikovoi, 1876–1899.* Edited by A. E. Gaisinovich and B. V. Levshin. Moscow: Nauka, 1978.

***Letters 2*:** I. I. Mechnikov. *Pis'ma k Ol'ge Mechnikovoi, 1900–1914.* Edited by A. E. Gaisinovich and B. V. Levshin. Moscow: Nauka, 1980.

***SV*:** I. I. Mechnikov. *Stranitsy vospominanii.* Moscow: USSR Academy of Sciences, 1946.

***ZIIM*:** O. N. Mechnikova. *Zhizn' Il'i Il'icha Mechnikova.* Moscow-Leningrad: Gosizdat, 1926.

ARCHIVES

AIP: Archives de l'Institut Pasteur, Paris
AIMR: Archives institutionnelles du musée Rodin, Paris
ARAN: Arkhiv Rossiiskoi Akademii Nauk, Moscow
DAHO: Derzhavnii Arkhiv Kharkivskoi Oblasti, Kharkov
DAKO: Derzhavnii Arkhiv Kievskoii Oblasti, Kiev
GARF: Gosudarstvennyi Arkhiv Rossiiskoi Federatsii, Moscow
RAC: Rockefeller Archive Center, Sleepy Hollow, NY
RGASPI: Rossiiskii Gosudarstvennyi Arkhiv Sotsial'no-Politicheskoi Istorii, Moscow
RGIA: Rossiiskii Gosudarstvennyi Istoricheskii Arkhiv, St. Petersburg
Saada Collection: The personal archive of Jacques Saada
The Waller Manuscript Collection, Uppsala University Library
Waldemar Mordechai Haffkine archive, National Library of Israel, Jerusalem

PART I: MY METCHNIKOFF

Chapter 1: Reversal of Fortune

the term *gerontology*: History professor W. Andrew Achenbaum, author of *Crossing Frontiers: Gerontology Emerges as a Science* (Cambridge: Cambridge University Press, 1995), has confirmed that Metchnikoff's use of *gerontology* was the first coinage of this term he had found (e-mail to author, August 17, 2010).
In a 1911 poll: "Who Are the Ten," *Strand Magazine.*
"Remember your promise": *ZIIM,* 218.
"a model of selfless service": Berkhin and Fedosov, *Istoriia SSSR,* 105.
"Leninism, the highest achievement": Ibid., 104.
authored the modern concept of immunity: Tauber, "The Birth of Immunology, III," 522; Tauber and Chernyak, *Metchnikoff and the Origins,* 135.
"an oriental fairy tale": EM, "Geschichte," repr., *ASS,* vol. 7, 504.

Chapter 2: The Paris Obsession

Lili had been "widely regarded": The first to mention this in print was Cavaillon, "The Historical Milestones," 414.
"He told it to you like a secret": Annick Perrot, interview with author, December 13, 2012.
"Biographies of great people should not cover up": EM, *Osnovateli,* repr., *ASS,* vol. 14, 226.

PART II: THE MESSINA "EPIPHANY"

Chapter 3: Eureka!

"Overall, Messina hardly stands out": EM, "Moe prebyvanie v Messine," repr., *SV*, 71.

had their greatest insights: See, for example, Johnson, *Where Good Ideas*, 108–13.

"It struck me that": *SV*, 74.

"Sensing that my hunch": Ibid.

"No sooner said than done": Ibid., 75.

the very beginning of 1883: The Metchnikoffs might have celebrated Christmas on January 6, according to the Russian Old Style calendar.

"Naturally, I was agitated": *SV*, 75.

Darwin had been right: For the development of Metchnikoff's views on Darwinism, see Alfred I. Tauber, "Introduction," in Gourko, Williamson, and Tauber, *The Evolutionary Biology Papers*, 1–21; Tauber and Chernyak, *Metchnikoff and the Origins*, 68–100; and Ghiselin and Groeben, "Elias Metschnikoff."

"Until then a zoologist, I suddenly became": *SV*, 75.

"It was like that blinding light": de Kruif, *Microbe Hunters*, 206.

In truth, his searches: EM's embryology research as a foundation for his phagocyte theory is analyzed in Tauber and Chernyak, *Metchnikoff and the Origins*, 25–67.

Chapter 4: A Boy in a Hurry

born on May 15, 1845: Kharkov metrics (DAHO 40.110.629, 1232-over) list Metchnikoff's date of birth as May 3, according to the Old Style Russian calendar, or May 15 by the New Style. Metchnikoff himself mistakenly stated his own birthday as May 16. The source of the mistake: he added thirteen days to the Old Style date, as was done in the twentieth century, but since he was born in the nineteenth, only twelve days should have been added.

spent his childhood in Panassovka: In all likelihood, Metchnikoff was born in the nearby village of Ivanovka. The metrics list Ivanovka as the place of residence of Metchnikoff's father (they make no mention of the mother). According to OM, the family moved to Panassovka shortly after Ilya's birth.

paid them two kopecks each: Bardakh, "Il'ia Il'ich," 1199.

Moldavian adventurer Nicholas Milescu: EM's forebear Nicholas Milescu (1636–1708), known as the Great Spatar, was a noted scholar, diplomat, translator, and author. People used to consult him instead of an encyclopedia, just as was to happen several generations later to the biologist Elie Metchnikoff. Selected sources for the Metchnikoff clan: *ZIIM*, 19–21; Spafarii, *Sibir'*, 3–13; "Rod Mechnikova," DAHO 14.11.5, 118-over-120.

Great Spatar: According to the encyclopedia *Sovetskaia Moldaviia* (Kishinev, 1982, pp. 74 and 587), "spatar"—from *spadă are*, the Romanian for "having a sword"—can have two meanings: a military commander in medieval Moldavia or an arms keeper who carried the ruler's sword at official ceremonies. The genealogical file on the Metchnikoff clan (DAHO 14.11.5, 118-over-120) shows that apart from Milescu, there was another Great Spatar in the Metchnikoff clan: Georgii Stepanovich, apparently Milescu's nephew, the grandfather of EM's great-grandfather Ilya.

The name Metchnikoff: The word *mechnik,* no longer in use in contemporary Russian, had several historic meanings: a court official attending "trials by iron" to probe a suspect's guilt, or a person carrying a sword in front of the monarch as a sign of royal power. Metchnikoff's name, however, is probably unrelated to these *mechniks,* but rather derived directly from *mech,* "sword."

Leiba Nevakhovich: Nevakhovich (1776–1831) was the first to spread among Russian Jews the ideas of a movement known as Haskalah, the so-called Jewish Enlightenment, which encouraged Jews to obtain secular education. In 1803, he published the first Russian-language book by a Jewish author, *Vopl' Dshcheri Iudeiskoi,* "Wailing of a Daughter of Judah," in which he called for the humane treatment of Jews and a relief of their suffering. In 1806, he obtained his own relief by converting to Christianity through the Lutheran Church. He moved in literary circles, meeting Alexander Pushkin, and wrote plays for the imperial theater. Later on, he returned for a while to the Russian part of Poland, where he subcontracted a monopoly on tobacco trade for the entire region and served in the Ministry of Finance in Warsaw. Selected sources: *Evreiskaia Entsiklopediia* (St. Petersburg: Obshchestvo dlia Nauchnykh Evreiskikh Izdanii, 1908–1913), 622–624, s.v. "Nevakhovich L. N."; Gessen, *Evrei,* 78–139; Kandel, *Kniga vremen,* vol. 1, 31–32; Kolodziejczyk, "Leon Newachowicz"; *Kratkaia Evreiskaia Entsiklopediia* (Jerusalem: Hebrew University of Jerusalem, 1990), 670; Nikolaev, *Russkie pisateli,* 244–45.

Catharine Michelson: The name of Metchnikoff's maternal grandmother, Catharine Michelson (1790–1837), appears in the inscription on her tombstone at the Lutheran Volkovskoe Cemetery in St. Petersburg, where she is buried next to her husband, Lev Nevakhovich (Gessen, *Evrei,* 138). She was most certainly a converted Jew like Nevakhovich (he was much more likely to have married a convert like himself than a Christian woman). Catharine's patronymic, Samoilovna, reveals that her father's name was Samuel. In all likelihood, she was the daughter of the Jewish painter and engraver Samuel Mikhailovich Michelson, the only Samuel Michelson listed in the St. Petersburg directory in 1809, around the time of Catharine's marriage to Nevakhovich. A Polish Jew like Nevakhovich, Michelson had first moved from Poland to Kiev, where he served as the leader of the Jewish community, and later to St. Petersburg (RGIA 535.1.7, 535.1.21, 789.1(I).1573, 1329.1.243 and 1374.4.243; DAKO 280.174.1060). Archival documents provide no complete listing of his family but they do supply indirect evidence that this Jewish painter was indeed EM's maternal great-grandfather. Thus, one 1838 document (RGIA 780.1(I).2272) reveals that Michelson had a younger daughter called Elena, who had converted to Christianity—probably the great-aunt Elena Samoilovna, mentioned in OM's book (*ZIIM,* 6). Yet another supporting detail is the name continuity in different generations of the family. As revealed by the painter Michelson's patronymic, his father had been called Mikhail, and so was one of Catharine's sons, EM's uncle Mikhail, brother of his mother Emilia.

half of the peasants still had no plows: Fedosov, *Istoriia SSSR,* 260.

more than three-quarters of the population was illiterate: Ibid., 393.

nearly two-thirds of the children died: Statistics for 1843. Gundobin, *Detskaia smertnost'*, 4.

messianic idea—then predominant among the Russian intelligentsia: Excellent sources include Pipes, *Russia under the Old Regime*, 249–280, and Fisher, "The Rise of the Radical."

"turned to science with enthusiasm": Bernstein, "Metchnikoff—'The Apostle.'"

"to have woken from a lethargic sleep": N. V. Shelgunov, quoted in Mikhail Lemke, *Ocherki po istorii russkoi tsenzury i zhurnalistiki XIX stoletiia* (St. Petersburg: Trud, 1904), 17.

***"Boga-net,"* "God-is-not":** *ZIIM*, 24.

"answer the major questions tormenting mankind": *SV*, 44.

"a weapon that can be used": Quoted in Antonovich and Eliseev, *Shestidesiatye gody*, 251.

"There's no doubt whatsoever": *SV*, 41.

"You are too sensitive": *ZIIM*, 33.

Chapter 5: Science and Marriage

"Darwin had fostered the hope": *SV*, 16.

Mendeleyev, who had just invented: Gordin, *A Well-Ordered Thing*, 22–30.

fictitious marriages: Kovalevskaia, *Vospominaniia*, 124; Bogdanovich, *Liubov' liudei.*

"A grave covered with flowers": *ZIIM*, 59.

"Why live?": *ZIIM*, 63.

"My Geneva catastrophe": EM to OM, April 6, 1898. *Letters 1*, 223.

"His thoughts turned to scientific issues": *ZIIM*, 63.

proposed by a German zoologist: Fritz Müller, in his 1864 booklet, *Facts and Arguments for Darwin.*

Only a most limited version of the theory: Mayr, *The Growth of Biological Thought*, 474–76.

"epoch-making": A. D. Nekrasov, "Raboty I. I. Mechnikova v oblasti embriologii," in *ASS*, vol. 3, 415.

"I delight in your energy": *SV*, 181.

He threw a tantrum: *ZIIM*, 43.

emptied a chamber-pot: Ibid., 50.

hurled a laboratory vessel at him: Antsiferov, *Iz dum o bylom*, 270.

Chapter 6: A Person of Extreme Convictions

Ilya jumped on one of the room's: Kondakov, "Vospominaniia," 6.

"had turned into quite a European city": Veniukov, *Iz vospominanii*, 3.

had even called him *"Mamasha"*: *ZIIM*, 45.

"I thought he looked rather like Christ": Ibid., 73.

"Despite all the excitement of the wedding": Ibid., 74; *Letters 1*, 54.

"I was thinking about my dear one": EM to OM, September 13, 1876. *Letters 1*, 27.

"Life shouldn't be valued": EM, *Etiudy optimizma*, 220.

"In the throngs of nervous angst": *ZIIM*, 85.

"After recovering from relapsing fever": Ibid., 86.

"Metchnikoff, a professor of zoology": Popovskii, *Vladimir Khavkin*, 18.

A Jewish lecturer in anatomy: Shtreikh, "Evreiskie otnosheniia." Whenever Metchnikoff encountered discrimination against Jews, he rushed to their defense, standing up for Jewish students and teachers at the university and, while managing his wife's estate, actively intervening on behalf of Jewish farmers who were persecuted by peasants and government alike.

"Attending council meetings became": *SV*, 81.

"The warden's betrayal": Ibid., 83; Shtreikh, "Iz professorskoi."

so did at least three other liberal professors: Reznik, *Mechnikov*, 173; Shtreikh, "Evreiskie otnosheniia," 9.

"The mere memory": *SV*, 229.

brought in 8,000 rubles a year: "Primechaniia," in *Letters 2*, 297.

Chapter 7: The True Story

Haeckel was a dilettante: EM, "Ocherk voprosa," repr., *ASS*, vol. 4, 321.

"more diligence than talent": Ibid., 320.

"unreliable, untimely, and unscientific": Ibid., 322.

"law" of creativity: Robinson, *Sudden Genius?*, 316–27.

When a starfish larva metamorphosed: EM, "Über die pathologische Bedeutung der intracellulären Verdauung," *Forschritte der Medicin* 2, no. 17 (1884): 558–69, repr., *ASS*, vol. 5, 45.

mobile cells ate up the surplus tissue: According to OM (*ZIIM*, 97), Metchnikoff at first hoped that the atrophy of surplus tissue was an instance of "physiological inflammation," one unrelated to disease. He then realized that such atrophy was not connected to the phenomenon he viewed as central to inflammation, that is, leukocytes exiting blood vessels, but was performed by cells located within the muscles.

Chapter 8: An Outsider's Advantage

"protection against disease is one of the most": EM, *Nevospriimchivost'*, 615.

the worst fear of Mithridates: William Smith, *A Dictionary of Greek and Roman Biography and Mythology* (London: John Murray, 1873), accessed September 11, 2015, www.perseus.tufts.edu/hopper/text?doc=Perseus:text:1999.04.0104:entry=mithridates-vi-bio-1.

left one-fifth of the world's population: Voltaire, "On Inoculation."

wise men in China: Needham with Gwei-Djen, *Science and Civilisation in China*, vol. 6, 155.

"The smallpox getting into the family": Voltaire, "On Inoculation."

Circassian baby girls had their tender skin: A. N. Sotin, *Zavisimost' ospiannykh epidemii ot ospoprivivaniia*, PhD diss., St. Petersburg, 1894, cited from Ulyankina, *Zarozhdenie immunologii*, 38.

"a woman of as fine a genius": Voltaire, "On Inoculation."

"highly intelligent and educated": EM, *Nevospriimchivost'*, 618.

Napoleon himself had a medal minted: Simmons, *Doctors and Discoveries*, 155.

"the merit and immense services": Ibid.

"merely communicates to animals": Ibid., 21.

one medical authority in Persia: Rhazes, *A Treatise on the Smallpox and Measles*, trans. W. A. Greenhill (London, Sydenham Society, 1848), cited from Silverstein, *Paul Ehrlich's Receptor Immunology*, 77.
a Renaissance scholar in Italy: Girolamo Fracastoro, *De Contagione et Contagionis Morbis et Eorum Curatione*, trans. W. C. Wright (New York: Putnam, 1930), 60–63, cited from Silverstein, *Paul Ehrlich's Receptor Immunology*, 78.
One eighteenth-century explanation: Silverstein and Bialasiewicz, "A History of Theories," 158.
(somewhat horticultural) theory: Silverstein, *Paul Ehrlich's Receptor Immunology*, 78.
"in some way incapable": *Comptes rendus de l'Académie des sciences* 90 (1880): 239.
Auguste Chauveau tried to shed light: *Comptes rendus de l'Académie des sciences* 89 (1880): 498; 90 (1880): 1526; 91 (1880): 148, cited from Silverstein and Bialasiewicz, "A History of Theories," 164.
"adaptation of the cells": EM, *Nevospriimchivost'*, 624.
the century's leading killer: Dubos and Dubos, *The White Plague*, 10.
a much more gradual account: EM, *Nevospriimchivost'*, 628–32.
a handful of researchers had actually suggested: In *Immunity*, Metchnikoff himself mentions several such suggestions, of which he was unware in Messina (*Nevospriimchivost'*, 627–28). They are also reviewed by Chernyak and Tauber, "The Birth of Immunology," 221–24, and by Ambrose, "The Osler Slide."
Working outside his own discipline: See also Tauber, "Ilya Metchnikoff."

Chapter 9: Eating Cells

"Das ist ein wahrer Hippokratische": *ZIIM*, 96.
"The body's nature is the physician": Hippocrates, *Volume VII. Epidemics.*
Virchow hurried to Messina: Virchow to Anton Dohrn, March 24, 1883. Groeben and Wenig, *Anton Dohrn und Rudolf Virchow*, 56.
"pope" and "pascha": Ackerknecht, *Rudolf Virchow*, 38; Simmons, *Doctors and Discoveries*, 11.
a fresh banner for drawing attention to cells: In 1885, Virchow was to publish an article, "The Struggle of Cells and Bacteria" (Ackerknecht, *Rudolf Virchow*, 114).
Etna never came through: Known eruptions took place in 1879 and 1886 (Groeben and Wenig, *Anton Dohrn und Rudolf Virchow*, 99).
"Recently, thanks to a lucky chance": *Berliner klinische Wochenschrift*, no. 324 (August 20, 1883): 526, quoted in "Primechaniia," in EM, *Pis'ma (1863–1916)*, 252.
"Pathologists believe and teach": EM, *Nevospriimchivost'*, 632.
"also plays a protective role": EM, "Untersuchungen über die intracelluläre Verdauung bei wirbellossen Tieren," *Arbeiten aus dem Zoologischen Instituten der Universität Wien* 5, no. 2 (1883): 141–68, repr., *ASS*, vol. 6, 20.
cells became *phagocytes*: *ZIIM*, 97.

Chapter 10: Curative Digestion

"The medical contingent of the congress": Bardakh, "Il'ia Il'ich," 1197.
"We often hear that so-and-so": EM, "O tselebnykh silakh," repr., *ASS*, vol. 13, 125–32.

"During the congress and for a long while": Bardakh, "Il'ia Il'ich," 1197.

"brought about a new era": Reznik, *Mechnikov*, 186.

"When this theory was attacked": EM, "Die Lehre von den Phagocyten," repr., *ASS*, vol. 7, 450.

"We now have more reason": EM, "Über die Beziehung der Phagocyten zu Milzbrandbacillen," *Virchow's Archive* 97, no. 3 (1884): 502–26, repr., *ASS*, vol. 6, 45.

"The war was launched": EM, *Nevospriimchivost'*, 634.

"which shows that the living organism": Baumgarten, "Referate," cited from *ASS*, vol. 6, 80.

"the limit of their life's duration": Baumgarten, *Lehrbuch der pathologischen Mykologie*, 1886, 1:114, quoted in *ASS*, vol. 6, 98.

"How wonderful that the spleen": Kleinenberg to EM, November 23, 1887. ARAN 584.4.142.

"You shouldn't be distressed": Kleinenberg to EM, April 8, 1885. Ibid.

Chapter 11: The Pasteur Boom

"Bound and howling": Duclaux, *Pasteur*, 294.

A CURE FOR HYDROPHOBIA: *New York Times*, October 30, 1885.

"an exceedingly important center": *Vedomosti Odesskago Gorodskago Obshchestvennago Upravleniia*, no. 10, February 1, 1886, 2.

"our famous scientist I. I. Metchnikoff": *Vedomosti*, no. 35, May 17, 1886, 4.

by the end of 1885, three had died: "L''affair Rouyer,'" *Pour la Science*, no. 33, November 2007–January 2008, 95.

***murderer*:** Ibid., 96.

"We've never had as many": Dr. Peter at the July 12, 1887, meeting. *Bulletin de l'Académie de médecine* 18 (1887): 38.

THE NEWARK CHILDREN: *New York Times*, December 22, 1885.

***"The Pasteur Boom"*:** *Puck*, December 23, 1885, quoted in Hansen, "America's First," 390.

"Nineteen people bitten": The story of the Smolensk bitten is told in Shevelev and Nikolaeva, *Poslednii podvig*, 32–41, 45–54.

MUZHIKS IN PARIS: *Le Matin*, March 15, 1886, 1.

Chapter 12: An Oriental Fairy Tale

"to provide accurate information": *Odesskii Listok*, June 13, 1886, 2.

free tickets granted them: EM, "Vospominaniia o Vitte," *Russkoe Slovo*, February 28, 1916, 3.

"Believe it or not": *Odesskii Listok*, June 21, 1886, 2.

"avoiding Pasteur's method": Reprinted in *Odesskii Listok*, October 20, 1886, 3.

"He tried to deduce from my gait": Bardakh, "Il'ia Il'ich," 1199.

"Can Pasteur's emulsion cause rabies": "Protokol zasedaniia Obshchestva odesskikh vrachei," no. 3, November 1, 1886, in *Oglavlenie Protokolov: God VII, 1886–87* (Odessa: Russkaia Tipografiia Isakovicha, 1888), 10.

"genuine persecution": *SV*, 84.

declared the meeting's protocol unprintable: Reznik, *Mechnikov*, 222.

"Who can conduct fruitful scientific research": "Deiatel'nost' Odesskoi bakteriologicheskoi stantsii," *Sbornik Khersonskago Zemstva* 20, no. 3 (May–June 1887), 78.

"The larger and less mobile macrophages": EM, "Über den Kampf der Zellen gegen Erysipelkokken: Ein Betrag zur Phagocytenlehre," *Virchow's Archive* 107, no. 2 (1887): 209–49, repr., *ASS*, vol. 6, 72.

"I had already expressed my opinion": EM, "Über die phagocytäre rolle der Tuberkelriesenzellen," *Virchow's Archive* 113, no. 1 (1888): 63–94, repr., *ASS*, vol. 6, 130.

"Metchnikoff's explanation of leukocyte activity": Baumgarten, *Zeitschrift für klinische Medicin* 15, no. 1 (1888): 1, quoted in EM, *Nevospriimchivost'*, 635.

"Metchnikoff's studies have brought no new observations": Ziegler, *Lehrbuch der allgemeinen pathologischen*, 341.

"An oriental fairy tale": EM, "Geschichte," 124, repr., *ASS*, vol. 7, 504.

"obstacles from above": EM, "Rasskaz o tom," repr., *SV*, 86.

"demanded practical results": Ibid., 84.

Chapter 13: A Fateful Detour

Sixth International Congress for Hygiene and Demography: "For" was replaced with "of" in the names of subsequent meetings on this topic.

"Hygienic piety has been rewarded": "The International Hygienic Congress," *British Medical Journal* 2, no. 1396 (1887): 739.

"spreading hygienic principles": Ibid.

"a peaceful little university town": *ZIIM*, 106.

most original and so creative: Louis Pasteur, "Lettre de M. Pasteur sur la rage," *Annales* 1, no. 1 (1887): 10.

"the sight of that parchment": Silverstein, *A History of Immunology*, 31.

"beer of revenge": Baxter, "Louis Pasteur's Beer."

"I saw a frail elderly man": EM, *Osnovateli*, 208.

"Even though my young colleagues": Ibid.

"I was in for a great embarrassment": Ibid., 211.

"We went to Strasbourg": *ZIIM*, 107.

Chapter 14: Farewell

"They were sure I was going": EM, *Osnovateli*, 227.

"Naturally, I ignored the warning": Ibid.

"My feeling of extraordinary respect": Ibid.

"crusty and opinionated tyrant": Brock, *Robert Koch*, 4.

"The man whom I saw": EM, *Osnovateli*, 228.

"Koch's response and his entire manner": Ibid.

"You know, I'm no expert": Ibid.

bitter turf disputes with the radical Virchow: Brock, *Robert Koch*, 82; Ackerknecht, *Rudolf Virchow*, 106–8.

nominating him for the very first Nobel Prize: Nomination Database. *Nobelprize.org*, accessed July 16, 2015, www.nobelprize.org/nomination/archive/. Metchnikoff later refused to support any other candidate until Koch would receive the prize (EM, *Pis'ma*, 188).

character assassination: Ackerknecht (*Rudolf Virchow*, 115) refers to this "character assassination" on the basis of unpublished letters from Koch to Carl Fluegge, in the archive of the Institute of the History of Medicine, Johns Hopkins University. Unfortunately, these letters have been lost. Johns Hopkins archivists have been unable to locate them.

"The contrast between our impression": *ZIIM*, 108.

"On the fourth day of the vaccinations": "Odessa. Podrobnosti neudachnoi privivki sibirskoi iazvy," *Moskovskie Vedomosti*, September 1, 1888, 5.

A veterinarian who studied the case: Vyacheslav A. Kuznetsov, "Professor Yakov Yulievich Bardakh (1857–1929): Pioneer of Bacteriological Research in Russia and Ukraine," *Journal of Medical Biography* 22, no. 3 (2014): 137.

"This painful episode": *ZIIM*, 108.

PART III: THE IMMUNITY WAR

The basis for my research into Metchnikoff's "immunity war" was Arthur Silverstein's "Cellular vs Humoral Immunity" (in his *A History of Immunology*, 25–42), which sets the early immunity debate in the historic context of the Franco-German relations. An analysis of the debate is provided by Tauber and Chernyak in "The Birth of Immunology," 447–73, and by the same authors in *Metchnikoff and the Origins of Immunology*, 154–74.

Chapter 15: The Temple on Rue Dutot

late nineteenth-century Paris: Selected sources: Benjamin, "Paris, Capital of the Nineteenth," 146–62; Jones, *Biography of a City*; Sowerine, *France since 1870*; Suzuki, *The Paris of Toulouse-Lautrec*, 18–25.

"useless and monstrous": "Au jour le jour. Les artistes contre la tour Eiffel," *Le Temps*, February 14, 1887, 2.

"whose clothes, made by an anonymous tailor": Joanne, *Paris illustré*, 64.

unthinkable 2.5 million francs: Jacques Mery, *Histoire des legs à L'Institut Pasteur*, Pasteur Institute brochure.

"with the whole world": *Times* (London), November 24, 1886.

"A few eloquent speeches": *SV*, 108.

its five research departments: Delaunay, *L'Institut Pasteur*, 51–60.

as a key to understanding evolution: "Morphology deals first of all with establishing the genealogy of the organic world," Metchnikoff wrote in his 1976 "Essay on the Question of the Origin of Species" (EM, "Ocherk voprosa," repr., *ASS*, vol. 4, 326).

"In Paris I finally realized": EM, "Rasskaz o tom," repr., *SV*, 86.

two hostile colossi, France and Germany: Selected sources: Bainville, *Histoire de deux peuples*; Binoche, *Histoire des relations franco-allemandes*; Howard, *The Franco-Prussian War*; Poidevin and Bariety, *Les relations franco-allemandes*; Tuchman, *The Guns of August*; Wawro, *The Franco-Prussian War*.

"like God in France": Tuchman, *The Guns of August*, 31.

dragged into a nationalist feud: Arthur Silverstein writes in *A History of Immunology* (p. 32): "We may only wonder whether this debate would have been as vitriolic or protracted, had the international political setting been different during the latter half of the nineteenth century."

critics were to hail from patriotic Prussia: Silverstein, *A History of Immunology*, 30.

"we, in Germany, do not quite accept": Rudolf Abel, *Centralblatt für Bakteriologie* 20 (November 15, 1896): 766.

"It's a grave mistake to adopt a nationalist": EM, *Nevospriimchivost'*, 652.

"Like dark clouds": *ZIIM*, 120.

Chapter 16: Engine in the Dark

"Most discussed at present": Behring, "On the Cause of the Immunity," 443.

"Not satisfied with destroying": EM, *Nevospriimchivost'*, 638.

"My experiments have shown": EM, "Le charbon des rats blancs," *Annales* 4, no. 4 (1890): 200.

Exposition Universelle: "The Great French Show," *New York Times*, May 19, 1889; Annegret Fauser, *Musical Encounters at the 1889 Paris World's Fair* (Rochester, NY: University of Rochester Press, 2005); Gaillard, *Paris: Les Expositions*, 46–69.

Gauguin and his artist friends: Thomson, *The Post-Impressionists*, 37–48.

"a mistake of nature": Metchnikoff cites Emmerich's comment in "La théorie des phagocytes au Congrès," 537.

"The phagocyte theory assumes": Carl Fränkel, *Grundriss der Bakterienkunde*, 3rd ed. (Berlin: Hirschwald, 1890), 203, quoted in *ASS*, vol. 5, 211.

"the phagocytic elements play an active": EM, "Sur la lutte des cellules de l'organisme contre l'invasion des microbes," *Annales* 1, no. 7 (1887): 327.

"the biochemical disfavor of the medium": Baumgarten, "Über das 'Experimentum crusis,'" 9.

"The phagocytes appear": *Zeitschrift für Hygiene* 4 (1888): 227, quoted in EM, *Nevospriimchivost'*, 639.

"Using this method": EM, "Le charbon des pigeons," *Annales* 4, no. 2 (1890): 80.

"unsatisfactory observations": Ibid., 75.

Chapter 17: An Amazing Friend

The physician Émile Roux: Selected sources: Cressac, *Le Docteur Roux*; Lagrange, *Monsieur Roux*; Perrot and Schwartz, *Pasteur et ses lieutenants*, 84–95.

"an avalanche of *Rs*": This apt description—which could apply to Metchnikoff or anyone else speaking with a Russian accent—is from Louis-Ferdinand Céline, *Journey to the End of the Night* (trans. Ralph Manheim, New York: New Directions Books, 2006, 243). Céline, who spent some time at the Pasteur Institute shortly after Metchnikoff's death, draws a vicious portrait of a Russian scientist, Serge Parapine, which must have been vaguely inspired by stories about Metchnikoff but, apart from the accent, is too grotesque to be taken seriously.

"Roux is practically the only person": EM to OM, May 9, 1893. *Letters 1,* 164.

"immediately fell under the spell": Lagrange, *Monsieur Roux,* 179.

"The Metchnikoffs provided": Ibid.

"this hardened bachelor": Roux had apparently been married briefly in England in his youth, but it seems that even his family didn't know about this (McIntyre, "The Marriage").

"My dear Papa Metchnikoff": For example, Roux to EM, November 27, 1910. Roux, *Pis'ma,* 148.

"the weather was magnificent": EM to OM, June 30, 1891. *Letters 1,* 113.

"That's yet another fruit": EM to OM, July 19, 1892. *Letters 1,* 147.

"As soon as he saw your handwriting": EM to OM, May 13, 1893. *Letters 1,* 167.

"When you come back": EM to OM, June 30, 1891. *Letters 1,* 113.

"When I saw that she not only wanted passionate love": Chernyshevsky, *What Is to Be Done?,* 320.

"I'm so happy when you talk about Roux": EM to OM, July 9, 1892. *Letters 1,* 138.

"At a certain time, Ilya Ilyich believed that my happiness": *ZIIM,* 77.

Metchnikoff used his privilege to nominate: In 1909. Nomination Database. *Nobelprize.org,* accessed July 17, 2015, www.nobelprize.org/nomination/archive/.

"I've always known that Roux": *ZIIM,* 206.

Chapter 18: Verdict

"His full energy as a scientist": *ZIIM,* 117.

"Only his intimates knew": Ibid., 120.

"Ever since our arrival in Paris": EM to G. N. Gabrichevskii, December 21, 1893. EM, *Pis'ma,* 138.

"When a German book or brochure": EM, *Osnovateli,* 217.

He mercilessly tore open the journals' covers: Besredka, "Vospominaniia," 40.

"the verdict of an international jury": EM, "La théorie des phagocytes au Congrès," 534.

to be handed down in Berlin in his absence: The English version of Metchnikoff's *Immunity* says that he himself presented his theory in Berlin—a mistranslation of the impersonal statement in the original French: "Le Congrès international de Médicine, réuni à Berlin en 1890, a été le premier où l'on ait parlé publiquement des nouvelles theories de l'immunité."

Learning that at the very last moment: Roux to the Metchnikoffs, July 30, 1890. Roux, *Pis'ma,* 37.

Metchnikoff implored Haffkine: OM to Haffkine, July 8, 1890. The Waller Manuscript Collection, Uppsala University Library, accessed October 8, 2015, www.theeuropeanlibrary.org/tel4/collection/a1059?id=a1059.

scouring it for reports from the Congress: EM to OM, July 30, 1890. *Letters 1,* 105.

six thousand or so "medical men": "Tenth International Medical Congress," *British Medical Journal* 2, no. 1545 (1890): 358.

"they were not morally at liberty": Ibid., 2, no. 1546 (1890): 415.

"As one entered the hall": Ibid., 2, no. 1545 (1890): 356.

"genuine supporters of peace": Rudolf Virchow, "Rede zur Eröffnung des X internationalen medicinischen Congresses," *Berliner klinische Wochenschrift* 32 (August 11, 1890): 724.

"For the first few days I could not rest": Joseph Lister to his brother, August 23, 1890. Godlee, *Lord Lister*, 511. In the same letter, Lister wrote about his own speech: "It was delivered to a kind audience, and I only hope it may do good rather than harm."

"It has long been very evident": Lister, "An Address on the Present Position," 378.

"The great Koch could do no wrong": Brock, *Robert Koch*, 292.

"Much ardor has been shown": Koch, "An Address on Bacteriological Research," 382.

Roux tried to soothe him: Roux wrote to EM on September 4, 1890: "I understand from your letter that you are worried about Koch's appraisal of the phagocyte theory, and that you are greatly upset. Mr. Koch is entirely incompetent in this matter." (Roux, *Pis'ma*, 39.)

Koch's Great Triumph: *New York Times*, November 16, 1890.

premature announcement under political pressure: Silverstein, *Paul Ehrlich's Receptor Immunology*, 10.

"cure," called *tuberculin*: For the story of tuberculin, see Goetz, *The Remedy*.

"the whole thing was experimental": *London Daily Telegraph*, November 20, 1890, 3, quoted in Markel, "The Medical Detectives," 2427.

Chapter 19: The Soul of Inflammation

***"Blut ist ein ganz besonderer Saft"*:** Behring and Kitasato, "Über das Zustandekommen," 454.

antibodies, at first called *antitoxins*: The term *antibody* was first used by Paul Ehrlich in 1891, but it took a few years for it to become widely accepted. See Lindenmann, "Origin of the Terms."

"I suspect that before many weeks": Lister, "Lecture on Koch's Treatment," 1259.

"When we received the report": EM and Émile Roux, letter to the editor (in French), *Deutsche Medizinische Wochenschrift*, no. 11, May 12, 1914, repr., *ASS*, vol. 14, 135.

"biological theory of inflammation": EM, *Leçons sur la pathologie*, x.

"a genealogical tree of inflammation": Ibid., 121.

"By means of the blood-current": Ibid., 220.

"the leukocytes themselves proceed": Ibid., 219.

"based on the law of evolution": Ibid., 230.

"I have indeed dared": EM, *Lectures on the Comparative*, xv.

"helped to set the stage for our modern understanding": Silverstein, "Introduction to the Dover Edition," in ibid., vi.

"Metchnikoff's policemen": Stefater et al., "Metchnikoff's Policemen."

"the extension of the normal to the pathological": Tauber, "The Birth of Immunology, III," 508–10.

"The curative force of nature": EM, *Leçons sur la pathologie*, 231.

***"generis humani saluti novum"*:** *Cambridge University Reporter* (June 23, 1891): 1104.

"a renewal of the objections to the theory": EM, *Lectures on the Comparative*, xv.

"not even permissible as a poetical conception": Review of "Entzündung" by Ernst Ziegler, *Real-encyclopädie der gesammten Heilkunde,* 3rd ed. (Vienna: Urban und Schwarzenberg, 1896), *British Medical Journal* 1, no. 1846 (1896): 1209.

Chapter 20: Under the Sword of Damocles

"I'm pretty nervous": EM to OM, August 11, 1891. *Letters 1,* 126.

"from every important country": "Congress of Hygiene and Demography," *British Medical Journal* 2, no. 1598 (1891): 387.

"the health of London": "Presidential Address on Recent Progress in Sanitary Legislation," *Lancet* 138, no. 3546 (1891): 338.

"Metchnikoff was delighted that the congress": *ZIIM,* 122. The original proposed location had been St. Petersburg, but, as the *Lancet* explained (October 1, 1887: 683), "this was considered a long and expensive journey, to say nothing of the prospect of Siberia if some of the delegates speak too freely."

"Bacteriological workers are at the present": "Bacteriology at the Congress of Hygiene," *British Medical Journal* 2, no. 1599 (1891): 436.

for nearly an hour instead of the allotted fifteen minutes: Bardakh, "Il'ia Il'ich," 1200.

"Let us take a concrete example": *Transactions of the Seventh International Congress,* vol. 2, 179.

"In this confrontation": Ibid., 182.

"Metchnikoff is busy showing": *ZIIM,* 122.

"carried the large audience": Godlee, *Lord Lister,* 515.

"the octaves": Cope, *The Versatile Victorian.*

"Each dish, each sort of wine": *SV,* 115.

"Nothing of the sort ever took place": Ibid.

"He didn't show off luxury": Ibid., 114.

"I consider meeting you": EM to Lister, December 16, 1895. R. I. Belkin, "Iz perepiski Mechnikova s angliiskimi uchenymi," *Voprosy Istorii Estestvoznaniia i Tekhniki* 1 (1956): 274.

"your sincere admirer": Lister to EM, February 18, 1897. ARAN 584.4.165.

"In the wake of the London Congress": *ZIIM,* 122.

Chapter 21: Law of Life

the most "Russian" of all European capitals: On the success of Russian emigration to France, see Foshko, *France's Russian Moment.*

"Russia is à la mode": *Le Figaro,* September 17, 1888, 1.

Portraits of the tsar and tsarina: Eksteins, *Rites of Spring,* 49.

a secret agent followed Metchnikoff around: "Spravka III-go Deloproizvodstva, June 14, 1894," GARF 102.92.1129, 8 and 8-over.

Another secret report from Paris: "Doklad Ego Prevoskhoditel'stvu Gospodinu Direktoru Politsii, April 6/18, 1894," GARF 102.318.1, 198-over.

helped track down a general: Popovskii, *Vladimir Khavkin,* 26–8.

mother was "a Jewess": Ibid.; and EM's autobiographical essay in *Les Prix Nobel en 1908,* 59, repr., *ASS,* vol. 16, 297.

"I ascribe my love for science": Bernstein, "Metchnikoff—'The Apostle.'"
"Russia has lost many great talents": Ibid.
"they are Jews, so the doors of the institute": "Beseda s sotrudnikami zhurnala *Vestnik Evropy*," *Vestnik Evropy*, no. 6 (1913): 382–85, repr., *SV*, 166.
Over the years, about a hundred students: Petrov and Ulyankina, "Luchshe umeret'," 432.
had not yet created bacteriology departments: Omelianskii, "Razvitie estestvoznaniia," 132.
"He shouted at me and used coarse language": Peter Argoutinsky to Louis Pasteur, November 7, 1891. AIP, DR.COR.1.
Metchnikoff had to reassure: *SV*, 98.
"When this nation becomes well educated": Burnet, *Un Européen*, 24.
a thirty-page essay, "Law of Life": EM, "Zakon zhizni," repr., *ASS*, vol. 13, 133–58.
"Scientists can't tell useful knowledge from useless": Basinskii, *Lev Tolstoi*, 292.
"Seekers of truth have not only": *ASS*, vol. 13, 134.
"You don't need to be a zoologist": Ibid., 138.
"Excessive exercising of muscles": Ibid., 147.
"Everything is good as it leaves the hands": Rousseau, *Emile*, 37.
"All organisms apart from man": EM's notebook, September 19, 1868. ARAN 584.1.4, 116.
"progressive evolutionist": *ZIIM*, 81.
"A sharp divide characterizes all vital issues": *ASS*, vol. 13, 155.

Chapter 22: Building a Better Castle

a bipolar disorder: Linton, *Emil von Behring*, 14.
"a ham was conserved against putrefactive": Bäumler, *Paul Ehrlich*, 42.
"metaphysical speculation": Emil Behring, *Die praktische Ziele der Blutserumtherapie* (Leipzig, 1892), 66, quoted in A. E. Gaisinovich and A. K. Panfilova, "Epistoliarnoe nasledie I. I. Mechnikova," in EM, *Pis'ma*, 28.
"relied on mysterious powers of the living cell": EM, "La théorie des phagocytes au Congrès," 541.
"interfered with treatment efforts": Ibid., 540.
"Born theoreticians like Buchner and Metchnikoff": Behring's 1893 lecture to Charité Society in Berlin, quoted in Linton, *Emil von Behring*, 89.
"I continue to deal with sick people": Behring to EM, October 7, 1891. ARAN 584.4.70.
"I am convinced that you have made": EM to Behring, November 29, 1891. EM, *Pis'ma*, 132.
"They envisaged transferring immunity": Lagrange, *Monsieur Roux*, 153.
three liters of serum a month: Ibid.
Ehrlich felt that by maneuvering him: Bäumler, *Paul Ehrlich*, 56–60; Linton, *Emil von Behring*, 169–172; Marquardt, *Paul Ehrlich*, 30–33; Silverstein, *Paul Ehrlich's Receptor Immunology*, 50–51.
slashed mortality from diphtheria almost in half: Grundbacher, "Behring's Discovery," 188.

estimated to have saved more than a quarter-million lives: Paget, *Pasteur and after Pasteur,* 100.

commonly—and mistakenly—viewed as the first successful treatment: Linton, *Emil von Behring,* 7.

a French law on patents prohibited profiting: Hage and Mote, "Transformational Organizations," 20.

"You're young, you need the money": Chistovich, "Iz zapisok."

"highly qualified loser": Luttenberger, "Excellence and Chance," 225–38.

"an exclusively French undertaking": "La guérison du croup à l'Institut Pasteur," *Le Figaro,* September 6, 1894, 1.

"the linguistically gifted Russian": Zeiss and Bieling, *Behring,* 114. Metchnikoff had full command of German and French and a limited knowledge of English, as well as some Italian and Portuguese. He read Darwin's *Origin* in German translation. Burnet wrote about him (*Un Européen,* 18): "Excited, inspired, fire in his eyes, shaking his long hair, he held discussions in French, Russian, German, understanding several other languages he didn't speak but responding in one he did: he coached, he prophesized—a real Demon of Science."

He gave Roux the honor: Linton, *Emil von Behring,* 199.

Metchnikoff served as godfather: Ibid. and "Primechaniia" in EM, *Pis'ma,* 264.

The godfather of the sixth son: Linton, *Emil von Behring,* 199.

"*Travaillons*": Ibid., 381.

"The capacity of blood serum to kill bacteria": EM, "Zur Immunitätlehre," in *Verhandlungen der Deutsche Gesellschaft für innere Medicin: XI Kongress in Leipzig* (Munich, 1892), 282–89, repr., *ASS,* vol. 6, 273.

"It's much more likely that the antitoxins": EM, "Ocherk osnov ucheniia o seroterapii," *Russkii Arkhiv Patologii, Klinicheskoi Meditsiny i Bakteriologii* 2, no. 1 (1896): 111–19, repr., *ASS,* vol. 7, 165.

"Hence the need to learn more": EM, "Ocherk sovremennykh napravlenii v terapii infektsionnykh boleznei," *Yuzhno-Russkaia Meditsinskaia Gazeta* 1 (1892): 3–5, repr., *ASS,* vol. 9, 82.

"It is well known that pure science": EM, "La théorie des phagocytes au Congrès," 540.

Chapter 23: The Demon of Science

nearly three hundred thousand people died: Popovskii, *Vladimir Khavkin,* 39.

more than eight thousand lives in Hamburg: *British Medical Journal* 1, no. 1678 (1893): 429.

"I've been swallowing *toutes les cochonneries*": "The Nature and Causes of Cholera," *Lancet* 144, no. 3709 (1894): 761.

"I keep chasing after cholera cases": EM to OM, June 5, 1893. *Letters 1,* 191.

"The 'psychosis,' as he was to call it later": *ZIIM,* 126.

"The flora of the human stomach": EM, "Sur l'immunité et la réceptivité vis-à-vis du cholera intestinal," *Annales* 8, no. 8 (1894): 586.

"Metchnikoff has invented yet another theory": "The International Congress of Hygiene," *Times* (London), September 14, 1894.

political activism spilled into a wave: Maitron, *Le movement anarchiste,* 257.

"Just when the theory of phagocytosis": EM, *Nevospriimchivost'*, 648.
"In the cholera infection of guinea pigs": Pfeiffer, "Weitere Untersuchungen," 4.
"I was ready to rid myself of life": *ZIIM*, 187.
"The pitiful waste of this brainy": de Kruif, *Microbe Hunters*, 218.
"unbelievably posh, gilded to disgust": EM to OM, September 3, 1894. *Letters 1*, 194.
"Concluding our overview": EM, "L'état actuel de la question de l'immunité," *Annales* 8, no. 10 (1894): 720.
"The audience was apparently": EM to OM, September 3, 1894. *Letters 1*, 193.
"I can see you now at the Budapest Congress": Émile Roux, "Lettre de M. É. Roux: Jubilé du Professeur Élie Metchnikoff," *Annales* 29, no. 8 (1915): 360.
"hats were thrown to the ceiling": Evelyn Maxine Hammonds, *Childhood's Deadly Scourge: The Campaign to Control Diphtheria in New York City, 1880–1930* (Baltimore: Johns Hopkins University Press, 1999), 92, quoted in Linton, *Emil von Behring*, 182.

Chapter 24: Kicking Against the Goads

"Extracellular Destruction of Bacteria in the Organism": *Annales* 9, no. 6 (1895): 433–61.
"Monsieur Menshikoff": Waldemar Haffkine. Manuscript of an article about Pasteur for *Menorah*, December 20, 1922. Waldemar Mordechai Haffkine personal archive, National Library of Israel, ARC. Ms. Var. 325.84.
"It's tough to see this slow extinguishing": EM to OM, May 9, 1893. *Letters 1*, 165.
"A Review of Several Works": EM, "Revue de quelques travaux."
"Macabre thoughts pursued": *ZIIM*, 146.
sculpting classes with the famous Jean-Antoine Injalbert: Bénézit, *Dictionnaire critique*, 81; *Khudozhniki russkogo zarubezh'ia*, s.v. "Mechnikova."
"I don't even know if there's an opera in Paris": *Novosti i Birzhevaia Gazeta*, no. 249, September 10, 1899, 2.
Jules Bordet, a young, unassuming: Selected sources for Bordet: Beumer, "Jules Bordet"; Schmalstieg and Goldman, "Jules Bordet"; Delaunay, "Jules Bordet."
"dear and revered *maître*": Jules Bordet, "Les leucocytes et les propriétés actives du serum chez les vaccinés," *Annales* 9, no. 6 (1895): 506.
furiously scolding a new trainee: Levaditi, "Centenaire."
Paul Ehrlich was blond: Selected sources for Ehrlich: Bäumler, *Paul Ehrlich*; Marquardt, *Paul Ehrlich*; Silverstein, *Paul Ehrlich's Receptor Immunology*.
"a brilliant eccentric": S. Flexner and J. T. Flexner, *William Henry Welch and the Heroic Age of American Medicine* (New York: Viking Press, 1941), cited from Bäumler, *Paul Ehrlich*, 13.
Ehrlich's side chain theory might not have had: Cambrosio, Jacobi, and Keating, "Ehrlich's 'Beautiful Pictures.'"
"puerile graphic representations": Jules Bordet, *Traité de l'immunité dans les maladies infectueuses* (Paris: Masson, 1920), 504, quoted in Silverstein, *Paul Ehrlich's Receptor Immunology*, 85.
playing into the hands of the humoral camp: "It is largely the opposition of Bordet and other distinguished workers in the Pasteur Institute that has spurred us on in our experimental labors, and caused us to establish the [side chain] theory more firmly than ever," Ehrlich wrote in the preface to his *Studies on Immunity* (p. vii).

a large file labeled "Polemics": Silverstein, *Paul Ehrlich's Receptor Immunology*, 141–2.
"I witnessed the concern and anxiety": Chistovich, "Pamiati L. A. Tarasevicha," 345.
"Antispermotoxin is not produced": EM, "Sur la spermotoxine et l'antispermotoxine," *Annales* 14, no. 1 (1900): 8.
"Ehrlich's theory does not at all contradict": EM, *Nevospriimchivost'*, 367.
"Both exceptionally lively": Bardakh, "Il'ia Il'ich," 1201.
"real *spiritus rector*": Ehrlich to Carl Salomonsen, February 24, 1899. RAC, Paul Ehrlich Collection, 650 Eh 89, copy book III, box 5, p. 13 (Hirsch translation, box 55, folder 2).
"will find it difficult to kick against the goads": Ehrlich to Thorvald Madsen, undated (November 5, 1898–February 21, 1899). RAC, 650 Eh 89, copy book II, box 80, folder 1 (from box 4), pp. 61/62 (Hirsch translation, box 55, folder 1).
"entire Metchnikoffian villages": Ehrlich to Carl Weigert, February 16, 1899. RAC, 650 Eh 89, copy book II, box 80, folder 3 (from box 4), pp. 473–74 (Hirsch translation, box 55, folder 1).
"The very able investigations": Ehrlich, "Croonian Lecture," 425.

Chapter 25: A Romantic Chapter

the French had somberly called *fin de siècle*: Weber, *France, Fin de Siècle.*
showcasing the magic of *fée électricité*: Ibid., 71.
the nickname the "Necropolitain": Ibid., 70.
more than fifty million people visited: Gaillard, *Paris: Les Expositions*, 88.
so many equestrian figures: Hausser, *Paris au jour le jour*, 29.
one of these went to Olga Metchnikoff: *Exposition Universelle de 1900: Liste des récompenses distribuées aux exposants. Supplément annexe au* Journal Officiel *du 18 aout 1900*, Group II, Classe 9, August 18, 1900, p. 37 (Ministère de commerce, de l'industrie, des postes et des télégraphes).
exhibited two busts: *Exposition universelle de 1900: Catalogue officiel illustré de L'Exposition Décennale des Beaux-Arts de 1889 à 1900* (Paris: Imprimeries Lemercier, 1900).
organizing, together with friends, a society called Montparnasse: *Khudozhniki russkogo zarubezh'ia*, s.v. "Mechnikova"; Severiukhin and Leikind, *Khudozhniki russkoi emigratsii*, 324.
donated her works to an art auction: Mnukhin, *Russkoe zarubezh'e*, vol. 2, 555.
took part in the work of the "Hungry Friday" shelter: Ibid., 138.
"He was like Sybil on a tripod": *ZIIM*, 133.
"We arrive at the conclusion": EM, "Sur l'immunité," *XIII Congrès Internationale de Médecine, Paris 1900: Comptes rendus*, vol. 3 (Paris: Masson, 1900), 25.
"You have won a smashing victory": Roux to EM, August 10, 1900. Roux, *Pis'ma*, 101.
"Science is far from having said": EM, *Nevospriimchivost'*, 690.
"gives one the definite impression": Brieger, "Introduction," *xxvi.*
"reads as pleasantly as a good novel": "Reviews: Theories of Immunity," *British Medical Journal* 2, no. 2185 (1902): 1595.
"Your favorable attitude": EM to Bordet, January 28, 1902. AIP, BDJ.6.
"to express my heartiest and most sincere": EM to Behring, December 11, 1901. EM, *Pis'ma*, 171.

"To Metchnikoff, we owe the first serious attempt": "Obituary: Elie Metchnikoff," *Lancet*, 159.

provided a fertile ground for immunity research: See, for example, two contemporary comments: "Discoveries and theories on which the modern science of immunity is based are largely attributable to the work and inspiration of Ehrlich and Metchnikoff." ("Immunity and Blood Affinities," *British Medical Journal*, April 16, 1904, 906); "The only [explanations of immunity] that have withstood the test of time are those of Metchnikoff, representing the cellular theory, and of Ehrlich, representing the humoral theory." (Harold C. Ernst, *Modern Theories of Bacterial Immunity*, Boston: Publication Office of the Journal of Medical Research, 1903, 10).

"to Paul Ehrlich and Elie Metchnikoff": George Nuttall, *Blood Immunity and Blood Relationship* (Cambridge: Cambridge University Press, 1904).

"If ever there was a romantic chapter": Lister, "Presidential Address," 741.

"Phagocytosis has been a story of disappointment": Nikolai Gamale'ia, *Osnovy obshchei biologii* (1899), 113, quoted in L. A. Zil'ber, "Fagotsitarnaia teoriia I. I. Mechnikova," in EM, *Voprosy immuniteta* (Moscow: USSR Academy of Sciences, 1951), 662.

PART IV: NOT BY YOGURT ALONE

Chapter 26: Haunted

"undisciplined hair": Bianchon, "Elie Metchnikoff."

"big head, which seemed expressly fitted": Mackenna, "Dr. Metchnikoff."

"small, mischievous eyes": Basset, "Vive la Vie!"

"guttural, flexuous, lilting voice": Bianchon, "Les 'Essais optimistes.'"

"much like one of those Bohemian book hunters": Mackenna, "Dr. Metchnikoff."

a favorite expression, *Donc alors*: d'Hérelle, "Autobiographie," 169.

"He was more than a father to me": Etienne Burnet to Léopold Nègre, February 21, 1951. AIP, NGR.3.

when he learned that the watchmaker's wife: de Clap'ie, "Professor Il'ia Mechnikov," 61–2.

"Since I'm paying such attention": EM to OM, July 9, 1902. *Letters 2*, 104.

"The period between fifty and sixty-five years": *ZIIM*, 117.

"The intensity of his feelings": EM, *Etiudy optimizma*, 221.

"Observing in his organism": EM, *Etiudy o prirode*, 131.

In ancient Rome: Grmek, *On Ageing*, 45.

required three lambs: Ibid., 46.

"In biblical times, it was thought that contact with young girls": EM, *Etiudy optimizma*, 133.

"The breath of young girls contains:" Quoted in ibid., 134.

For those reading Genesis literally: Shapin and Martyn, "How to Live Forever," 1580.

All that changed in the eighteenth century: Gruman, *A History of Ideas*, 74–75.

"The rapid progress true science now makes": Franklin to Joseph Priestley, 1780, quoted in Gruman, *A History of Ideas*, 74.

dissected thousands of corpses and countless animals: Wendy Moore, *The Knife Man* (New York: Broadway Books, 2005), 4.

"I had imagined that it might be possible to prolong": The experiment was performed in 1766. John Hunter, *Lectures on the Principles of Surgery* (Philadelphia, 1841), 76, quoted in Gruman, *A History of Ideas*, 84.

Brown-Séquard, then seventy-two, unveiled: Aminoff, *Brown-Séquard*, 203–11; Stambler, *A History of Life-Extensionism*, 28–30.

"other forces": Charles-Édouard Brown-Séquard, "Des effets produits chez l'homme par des injections sous-cutanées d'un liquid retiré des testicules frais de cobaye et de chien," *Comptes rendus hebdomadaires des séances et mémoirs de la Société de biologie* 1, series 9 (1889): 418.

leading to hormone replacement therapy: Stambler, "The Unexpected Outcomes."

"It's really surprising that so few precise facts": EM, "Revue de quelques travaux," 261.

Chapter 27: *Vive la Vie!*

"*VIVE LA VIE!*": Basset, "Vive la Vie!"

"None of us should despair": Henri des Houx, "Les Pasteuriens," *Le Matin*, January 1, 1900, 1.

"Metchnikoff is really annoyed by the noise": Jules Bordet to Marthe Bordet-Levoz, January 5, 1900. AIP, BDJ.3.

"I've been a victim": Mackenna, "Dr. Metchnikoff."

"Aging is a disease that should be treated like any other": EM, *Etiudy o prirode*, 3.

"Senile degeneration can be reduced": EM, "Revue de quelques travaux," 263.

"In the free moments, my thoughts": EM to OM, September 10, 1901. *Letters 2*, 62.

"Scientists have been neglecting aging": EM, "Sur le blanchiment des cheveux et des poils," *Annales* 15, no. 12 (1901): 865.

Metchnikoff was not entirely off the mark: See, for example, Tobin and Paus, "Graying."

Queen Marie Antoinette: "Comment les cheveux blanchissent," *Le Temps*, January 10, 1902, 2.

women who curled their hair: "A travers Paris," *Le Figaro*, May 10, 1906, 1.

X-rays might be even more helpful: "A l'Académie des Sciences," *Le Petit Parisien*, July 17, 1906, 2.

Chapter 28: Law of Longevity

turning into a short-lived obsession: For an excellent history of the autointoxication theory, see Dally, *Fantasy Surgery*, 66–83.

having their patients swallow disinfecting mixtures: Rettger and Cheplin, *A Treatise on the Transformation*, 3.

"and this leads to premature aging of our tissues and organs": EM, "Sur la flore," 33.

"derive no benefit from this organ": Ibid.

Aristotle had noted that larger animals: Silvertown, *The Long and the Short*, Kindle edition, loc. 343.

longevity of living beings was said to be proportional: Stambler, *A History of Life-Extensionalism*, 12–15.

"I'm almost inclined to derive a general rule": EM, "Sur la flore," 32.
"the Englishmen delivered long speeches": EM to OM, April 23, 1901. *Letters 2*, 49.
"no age-related changes except that his character": EM, "Recherches sur la vieillesse des perroquets," *Annales* 16, no. 12 (1902): 913.
"Perhaps I'll suddenly take up philosophy": EM to OM, August 25, 1900. *Letters 2*, 45.

Chapter 29: The Nature of Man

"Our strong will to live runs counter": EM, *Etiudy o prirode*, 238.
***The Nature of Man*:** P. Chalmers Mitchell noted in his "Editor's Introduction" to the English edition (London: William Heinemann, 1903, p. viii) that "human constitution" sometimes better conveys the meaning, as in French, *la nature humaine* implies "not only the mental qualities of man, but his bodily framework."
"significantly above a hundred years": EM, *Etiudy o prirode*, 260.
"This instinct must be accompanied by marvelous sensations": Ibid., 265.
"They all yearned for a long life": Ibid., 134.
"This phrase, which sounds so strange to our ears": Ibid., 263.
"Old men, incapable of either inspiring or satisfying": Ibid., 107.
masturbation at a young age could "ruin health": Ibid.
"It is remarkable how enormously the disharmonies": Ibid., 106.
"the greatest living philosopher": Ibid., 271.
"For Herbert Spencer, the great complexity of life": Ibid., 273.
"Consequently, there is no blind striving": Ibid., 38.
"Human constitution is mutable": Ibid., 270.
"The purpose of human existence lies in": Ibid., 22
"merely living longer": EM, "Introduction," *Sur la fonction sexuelle*, repr., *ASS*, vol. 16, 283.
"Perhaps this belief in the incidence of death": Sigmund Freud, *Beyond the Pleasure Principle*, trans., C. J. M. Hubback, part 4 (London, Vienna: International Psycho-Analytical, 1922), accessed May 11, 2015, www.bartleby.com/276/6.html.
"some striking similarities": Cochrane, "Elie Metchnikoff and His Theory," 32–34.
"Freud wanted to be a scientist": Mark Aveline, "Commentary on Cochrane Al (1934)," *International Journal of Epidemiology* 32 (2003): 35.
"It is most likely that a scientific study of aging": EM, *Etiudy o prirode*, 278.
"people help one another much more": Ibid., 280.
"If it's true that one cannot live without faith": Ibid., 282.
"one of the most important scientific productions": "Metchnikoff on Human Nature," *New York Times*, June 20, 1903, BR15.
in Great Britain, where Metchnikoff's book was perhaps most enthusiastically: One famous fan was H. G. Wells; he was quoted by the *London Speaker* as calling the book "extremely interesting." Metchnikoff's message must have resonated with Wells's own ideal of entrusting humanity's future to a self-selected caste of experts guided by science.
"We will not quarrel with him for the paucity of authentic facts": "A Naturalist on Human Nature," *Times Literary Supplement*, 159.

***auf der Luft gegriffen*:** EM, "Bor'ba so starcheskim pererozhdeniem," *Russkie Vedomosti*, no. 130, June 7, 1912, repr., *ASS*, vol. 15, 346.
"One must also show how to achieve": EM, *Etiudy o prirode*, 22.

Chapter 30: Papa Boiled

"a disgusting and stinking habit": EM to OM, April 11, 1898. *Letters 1*, 215.
"An impeccably dressed saleswoman": ARAN 584.1.30, repr., *ASS*, vol. 10, 17–19.
"If one can't eat fruit": "La campagne contre le raisin et contre le vin," *La Justice*, October 1, 1906, 1.
"Monsieur *le professeur*, come buy my vegetables": "A Travers la Presse. D'*Excelsior*: Le 'père Bouilli,'" *Le Gaulois*, July 18, 1916, 3.
"intestinal putrefaction": Metchnikoff discusses the controversies over this issue in "Sur les microbes de la putréfaction intestinale," *Comptes rendus hebdomadaires des séances de l'Académie des sciences* 147 (1908): 579–81.
swallowed pieces of rotten horse meat: Tissier and Gasching, "Recherches sur la fermentation," 561.
"The milk microbes consistently produce": EM, *Etiudy o prirode*, 239.
The liquid, slightly alcoholic koumiss: For historic descriptions of fermented milks, see Douglas, *The Bacillus*.
"If a man cannot reconcile himself to sour milk": Duncan W. Freshfield, *The Exploration of the Caucasus*, 1896, quoted in Douglas, *The Bacillus*, 21.
"nourish persons when nothing else will": "Koumiss, a Real Milk Wine," *Times* (London), June 10, 1875, 15.
started drinking sour milk regularly in his early fifties: In his 1905 brochure on sour milk, Metchnikoff writes that he'd been drinking it for seven years.
"These findings explain why lactic acid often stops": EM, *Etiudy o prirode*, 240.
twenty-five grams per liter of milk: EM, *Etiudy optimizma*, 167.
Life expectancy in much of western Europe and North America had soared: In the United States, it went from 33 in 1800 to 49.7 in 1900 (Grmek, *On Ageing*, 32); in France, it went from 24.7 in 1740–1745 to 50.4 in 1909–1913 (Vallin, "Mortality in Europe," 47).
for older people, on average, mortality rates had dropped only slightly: Grmek, *On Ageing*, 32; Riley, *Rising Life Expectancy*, 1.
numbers of centenarians began to decrease: Grmek, *On Ageing*, 37.

Chapter 31: The Butter-Milk Craze

"*La Vieillesse*": EM, "La Vieillesse," repr., *ASS*, vol. 15, 34–56.
"To the casual stranger": Brandreth, "The Man Who Prolongs," 583.
"The man is possessed": Ibid.
"dogs capture seals": EM, "La Vieillesse," 37.
"The dog seemed offended by the comparison": G. Davenay, "Une conférence du Dr. Metchnikoff," *Le Figaro*, June 9, 1904, 3.
"It's a source of all sorts": EM, "La Vieillesse," 54.

"Interestingly, this microbe is found": Ibid., 53.

"Those of you, pretty ladies": "Pour supprimer la vieillesse," *Le Temps,* February 10, 1905, 3.

CAN OLD AGE BE CURED?: *Pall Mall Magazine,* no. 34, 1904, 205–09.

"any one desiring to attain a ripe old age": *Chicago Daily Tribune,* September 20, 1904, 1.

saw light in France in the fall of 1905: The publication of the French version was announced by *Le Figaro* on November 11, 1905.

several English translations: Worldcat.org listings include *Genuine Lacto-Bacilline: A Medicine and Food* (New York: Baldwin Hygienic Dairy Products Inc., 1906[?]); *Scientifically Soured Milk* (Lacto-Bacilline Co. of New York, May 1907); *Notes on Soured Milk and Other Methods of Administering Selected Lactic Germs in Intestinal Bacterio-therapy* (London: John Bale, Sons & Co., 1909); *Observations on Curdled Milk* (Cornell University Library, date unknown).

the brochure contained instructions for preparing: In a letter of January 27, 1907 (EM, *Pis'ma,* 197), Metchnikoff admonished a friend in Russia for asking whether he had a secret formula for preparing the sour milk: "You know only too well there is no 'patent' whatsoever, nor can there be."

"Its use at five hundred to seven hundred milliliters": EM, *Quelques Remarques sur le Lait Aigri,* repr., *ASS,* vol. 15, 262.

"Clearly, we do not look upon the milk microbes as an elixir": Ibid., 263.

He was preparing his own sour milk by placing: Roux to EM, October 3, 1905. Roux, *Pis'ma,* 144.

He used pure carbolic acid to discourage: John Harvey Kellogg, *Plain Facts for Old and Young* (Battle Creek, MI: Good Health Publishing Company, 1910), 326.

"placed the whole world under obligation": Kellogg, *Autointoxication,* 313.

"thus planting the protective germs": "Dr. John Harvey Kellogg," accessed January 29, 2012, www.museumofquackery.com/amquacks/kellogg.htm.

"On the Use of Soured Milk": Herschell, "On the Use of Selected," 371.

rinsing their vaginas with whey: For modern studies on yogurt consumption and vaginal infections, see David Drutz, "Lactobacillus Prophylaxis for *Candida* Vaginitis," in *Annals of Internal Medicine* 116, no. 5 (1992): 419–20.

"The milk was very sour, but this didn't prevent": EM, "Les microbes lactiques," repr., *ASS,* vol. 15, 273.

"since Metchnikoff began to publish his investigations": Mason, "Notes on the Lactic Ferments," 957.

"It may be well to direct the attention": R. Tanner Hewlett, "Sour Milk," letter to the editor, *Lancet* 175, no. 4514 (1910): 677.

"Yoghourt can be used for an indefinite time": "Curdled Milk and Intestinal Decomposition," *British Medical Journal* 2, no. 2543 (1909): 48.

"to taste the delicious Bulgarian curdled milk": *Le Figaro,* March 17 and 27, 1905.

"In mundane circles": "Letter from Paris," *Boston Medical and Surgical Journal* 158, no. 4 (1908): 140.

blossomed into an international business: See, for example, the following statement by Hastings and Hammer in "The Occurrence and Distribution of a Lactic Acid Organism": "Due to the ideas of Metchnikoff concerning the therapeutic value of milk fermented by this organism, fermented milks have been widely introduced into this country [the United States] and abroad."

"packed in a dainty chocolate cream": "Massolettes and Sour Milk," advertisement. *Times* (London), March 10, 1910, 9.

Metchnikoff constantly had to intervene: See, for example, his letter to D. N. Anuchkin in EM, *Pis'ma*, 201.

Metchnikoff had to appeal to a court: "Tribunal civil de la Seine," *Gazette du Palais*, March 4, 1905, 418.

Chapter 32: A True Malady

"a small oasis of science and art": Chistovich, *I. I. Mechnikov*, 16.

"If I weren't a scientist I would have been": Burnet, *Un Européen*, 26.

"How pure the air is": *ZIIM*, 118.

"I dedicate this book to my dear, beloved girl": The page with the dedication is reproduced in *Letters 2*, 109.

"He particularly loved one of his goddaughters": *ZIIM*, 199.

"have no children, but he has a godchild": Slosson, *Major Prophets*, 158.

It had begun in November 1894: EM wrote to Marie on November 12, 1904: "Today is the third anniversary of your marriage and the tenth anniversary of the day I started treating you for anemia."

She was twenty-two: Marie Rémy was born in October 1872; she died in 1964.

"When I examine a myxomycete": Burnet, *Un Européen*, 25.

Émile Rémy, two years her junior: Émile Rémy was born in July 1874; he died in 1944.

"to take care of our dear little Lili": EM to the Rémys, August 24, 1906. Saada Collection.

"The simplest phenomena": EM, *Etiudy optimizma*, 222.

"In other words, from a pessimist": Ibid., 223.

"He kissed my hands and cried": Saada-Rémy, "Souvenirs de Lili," 20.

"His love for children": *ZIIM*, 199.

one in four children died in his or her first year: EM, *Etiudy optimizma*, 84.

"My love for Lili is a true malady": EM to Marie Rémy, March 9, 1911. Saada Collection.

to sponsor a commercial enterprise, Société Le Ferment: *ZIIM*, 183–84; "L'histoire du lait caillé," in Delaunay, "A propos de la famille Saada," 4; EM's letters of December 6, 1904, and July 20, 1905, to *Le Ferment*, reprinted in *Novoe Vremia*, undated clipping, ARAN 584.1.334, 8.

concession to the London Pure Milk Association: "The Truth about Sour Milk," advertisement. *Times* (London), March 15, 1910, 14.

the Ferment Company of New York: Headquartered at 124-126 West 31st St., New York City. Letter of a representative of the Ferment Company to EM, May 15, 1909, ARAN 584.3.32; *Ferment Company of New York: Minute Book, 1907–1916*, Hagley Museum & Library, Wilmington, DE, online summary, accessed August 31, 2015, http://socialarchive.iath.virginia.edu/ark:/99166/w6jt6724.

"He answered that the choice had been": *ZIIM*, 184.

Chapter 33: Absurd Prejudice

Lili was punished for calling a schoolteacher: Saada-Rémy, "Souvenirs de Lili," 16.
"I protest against this exaggeration": EM to Marie Rémy, October, 20, 1914. Saada Collection.
"I live and breathe anew": EM to Marie Rémy, August 20, 1908. Saada Collection.
"I think about you every day": EM to Lili, August 20, 1908. Saada Collection.
"Tender kisses to my dear Marie": EM to Marie Rémy, February 5, 1911. Saada Collection.
the Colony: In Condé-sur-Vesgre. Inspired originally by Charles Fourier's idea of a *phalanstère* utopian community, it served as a getaway for Parisian artists, musicians, writers, and others. According to Danielle Duizabo (e-mail to author, December 1, 2012), Olga was a member from 1904 to 1910.
"beloved fugitive": EM to OM, September 17, 1901. *Letters 2*, 72.
"I love you very, very deeply": EM to OM, April 21, 1901. Ibid., 47.
"how groundless it is to think about separation": EM to OM, September 4, 1901. Ibid., 52.
"And what about your attachment to Roux?": EM to OM, July 26, 1906. Ibid., 198.
"It's hard for me, knowing that I'm returning home": Ibid.
"Of course I'm attached to the child": Ibid.
"Émile loved Marie more": Delaunay, "A propos de la famille Saada," 3.
"as if there was something about his family life": EM, *Osnovateli*, 226.
"the absurd prejudice, which today governs": Ibid.
"couldn't forgive him his scientific superiority": Ibid.
"With time, when marital relations": Ibid.
"just as there are bearded ladies": Bernstein, "Metchnikoff—'The Apostle.'"
"When I need a good dinner": *Tolstoi i Mechnikov o zhenshchinakh*, a brochure (St. Petersburg: Tipografiia Ulei, 1909[?]), 16.
"It would be very important for humanity's well-being": EM, "Vospominaniia o Sechenove," *Vestnik Evropy*, no. 5, (1915), 68–85, repr., *SV*, 61.
"Not only artistic creativity but other forms of genius": EM, *Etiudy optimizma*, 241.
"Scientists are no exception to this rule": Ibid.
"It would be most interesting to determine": Ibid., 245.
"since ancient times, sexual function had been closely associated": EM, *Sur la fonction sexuelle*, repr., *ASS*, vol. 16, 286.
"It's unacceptable for the modern": Ibid., 287.

Chapter 34: Return of a Psychosis

Historically, people in different countries had blamed: Frith, "Syphilis," 50.
***sifilizatsia* of the population:** *Rasprostranenie sifilisa v Rossii*, 2.
"A night with Venus and a lifetime with mercury": M. Dobson, *Disease* (London: Quercus, 2007), quoted in Frith, "Syphilis," 53.
cost an exorbitant 1,000 to 2,000 francs: EM to V. A. Morozova, a Moscow philanthropist, December 21, 1904. EM, *Pi'sma*, 188. Edwige cost 1,150 francs.
roughly half the annual salary: 1904 Pasteur Institute payroll, ARAN 584.3.7.

Moscow Prize: Created with donations made on the occasion of the Twelfth International Medical Congress held in Moscow in 1897. EM, *Pis'ma*, 271.
"a disease unknown in the forests": "Est-elle transmissible? L''avarie' chez les singes," *Le Matin*, July 29, 1903, 2.
"the only means for science": EM, *Etiudy optimizma*, 261.
Olga and friends organized the benevolent Montparnasse society: Severiukhin and Leikind, *Khudozhniki russkoi emigratsii*, 324.
When a Russian sculptor was unsure: EM's letter to Rodin, September 13, 1908, about two marble busts by Anna Golubkina (AIMR). On March 16, 1909, EM wrote to Rodin with a similar request, concerning the acceptance of three busts by Olga to the *Salon de la Societé Nationale*. In both cases, the works were accepted.
a young revolutionary had a problem: Zegebart, "Vospominaniia."
Russian Higher School of Social Sciences: For the story of the school, on 16 rue de la Sorbonne, see Gutnov, "L'École russe."
under the pseudonym Ilyin: "Perechen' kursov, prochitannykh v 1902–3 akademicheskom godu," RGASPI 4.1.124, 1–6.
"Every time I meet our friend Ilya Metchnikoff": Guriel, "Lenin and Weizmann."
total number of Russians there to about thirty-five thousand: This figure is for the year 1911. Nicolas Ross, *Saint-Alexandre-sur-Seine* (Paris: Le Cerf, 2005), 263, cited from Foshko, *France's Russian Moment*, 66.
"They are dressed very poorly": EM to OM, July 31, 1906. *Letters 2*, 205.
"Several Russian ladies came today": EM to OM, July 10, 1906. Ibid., 182.
"The discoverer of the phagocytes": "A Naturalist on Human Nature," *Times Literary Supplement*, 159.
"His laboratory at the Pasteur Institute was a salon": Burnet, *Un Européen*, 17.
Whenever a group of people gathered: Besredka, "Vospominaniia," 38–42.
"an orgy of interferences": EM to OM, July 20, 1910. *Letters 2*, 241.
"An indescribable number of Americans": EM to OM, September 14, 1907. Ibid., 224.
"He's completely pig-headed": EM to OM, July 13, 1906. Ibid., 186.
His lab was among many in the world: EM, "Recherches microbiologiques sur la syphilis," *Bulletin de l'Académie de médecine* 54 (1905): 468–76, repr., *ASS*, vol. 10, 262–28.
he confirmed the presence of the coiled microbe: Ibid., 265–67.
"Working under such conditions, no wonder": EM to OM, June 16, 1905. *Letters 2*, 160.
In monkeys and apes, such blocking proved possible: EM, "Études expérimentales sur la syphilis," *Annales* 19, no. 11 (1905): 673–98.
a mercury compound: This chloride of mercury was used to treat syphilis, among many other ills.
On February 1, 1906, in the presence: EM, "Recherches sur la syphilis," *Bulletin de l'Académie de médecine* (1906): 554–59, repr., *ASS*, vol. 10, 307–9.
"the hero who has risked his health": "La Thèse de l'Inoculé," *Le Matin*, July 21, 1906, 2.
"immoral to let people believe": EM, *Etiudy optimizma*, 260.
"Is it immoral to spread a means that rescues": "Entre Savants: Controverse sur l'Avarie," *Le Matin*, July 22, 1906, 1.
"Everything having to do with sexual function": EM, *Sur la fonction sexuelle*, repr., *ASS*, vol. 16, 284.
He further called for sex education in schools: Ibid. and EM, *The New Hygiene*, 79–80.
"He came to entirely revolutionary conclusions": *ZIIM*, 199.

For all the controversy, *la pommade de Metchnikoff*: In *Etiudy optimizma* (p. 141), Metchnikoff cites the statement of a German physician at a medical congress in Bern in 1906 (*Die experimentelle Syphilisforschung*, Berlin, 1906, 82): "It's our duty as physicians to strongly recommend the use of the 30-percent calomel ointment, tested by Metchnikoff and Roux, in all cases that might lead to a syphilis infection."

"Prevention of sexually transmitted": "Prophylaxie des maladies vénériennes dans l'armée," *La Presse Médicale*, November 10, 1907.

the most widely prescribed drug in the world: Amanda Yarnell, "Salvarsan," *Chemical & Engineering News* 83, no. 25 (June 20, 2005), accessed January 4, 2014, http://pubs.acs.org/cen/coverstory/83/8325/8325salvarsan.html.

Chapter 35: Biological Romances

Sir Almroth Wright: Selected sources for Wright: Dunnill, *The Plato of Praed Street*; Marvin J. Stone, "The Reserves of Life: William Osler versus Almroth Wright," *Journal of Medical Biography* 15, supplement 1 (2007): 30.

"No Sir, I have given you the facts": Silverstein, review of *The Plato of Praed Street*, 159.

"The blood fluids modify bacteria": A. E. Wright and Stewart R. Douglas, "An Experimental Investigation of the Role of the Blood Fluids in Connection with Phagocytosis," *Proceedings of the Royal Society of London* 72 (1903): 366, accessed September 3, 2015, doi:10.1098/rspl.1903.0062.

"The opsonic index affords the physician": Leonard Keene Hirshberg, "End of Disease," *New York Tribune*, April 9, 1911, SM5.

"Wright is the sort of man": "Sir Almroth Wright's Reply to Sir Henry Dale's Tribute," *British Medical Journal* 1, no. 3926 (1936): 707.

"far less important": EM, *The New Hygiene*, 23.

"The Englishmen were exceptionally gracious": EM to OM, May 26, 1906. *Letters 2*, 170.

"I was the only one in a frock coat": EM to OM, May 25, 1906, Ibid.

"wither your renown has long since preceded you": Edward A. Beck to Metchnikoff, May 19, 1906. ARAN 584.4.66.

"We'd be in poor shape, with our dealings": EM to OM, August 25, 1903. *Letters 2*, 121.

4,000 francs a year: The 1904 Pasteur Institute payroll, ARAN, 584.3.7. According to Dr. Doyen's letter to *Action Française* (ARAN 584.1.335, 22), EM's salary was later raised to 6,000 francs a year, and in 1910, after the Pasteur Institute received the Osiris donation, to 15,000 francs a year.

about a quarter of the income: OM's annual income from Russia, 8,000 rubles a year, was then equivalent to about 16,000 francs.

an assurance from a senior Russian politician: Witte, *Memoirs*, vol. 3, 355–6.

"Sir Almroth Wright, following up one of Metchnikoff's most suggestive": Shaw, *The Doctor's Dilemma*, accessed August 26, 2014, www.online-literature.com/george_bernard_shaw/doctors-dilemma/0/.

"If the phagocytes fail": Ibid.

"Despite the activity of opsonins": EM, "Bericht über die im Laufe les letzten Dezeniums erlangten Fortschritte in der Lehre über die Immunität," *Ergebnisse der Pathologie* 1 (1907): 645–89, repr., *ASS*, vol. 7, 356.

"These achievements of his have lately": Lagerkvist, *Pioneers of Microbiology*, 164.

Chapter 36: *Mais C'est Metchnikoff!*

"Tables raised from all four legs": Pierre Curie to Georges Gouy, July 24, 1905, quoted in Quinn, *Marie Curie,* 208.
"They all ended in failure": EM, *Sorok let,* repr., *ASS,* vol. 13, 24.
"to curtail human suffering": Ibid., 13.
fascination with spiritualism and the occult had soared: Weber, *France, Fin de Siècle,* 32–34.
"people of all ages, from youth of both sexes": EM, *Sorok let,* 29.
"Death is a very interesting experiment": Ibid., 35.
"Obviously, the prolongation of life should go hand in hand": EM, *Etiudy optimizma,* 132.
"a surprisingly relevant book today": Bial, "Optimism and Aging," 541.
"We can suppose humans can reach the age": EM, *Etiudy optimizma,* 87.
"We can conclude that decrepitude": Ibid., 29.
"The contest between optimism and pessimism": Shrimpton, "'Lane, You're a Perfect Pessimist'," 43.
"Added to gloominess transmitted to us": "Optimist Club Believes It Has Secret of Long Life," *New York Times,* February 25, 1912, SM7.
"People are capable of great deeds": EM, *Etiudy optimizma,* 284.
"by daringly planting himself in front of Old Age": Bianchon, "Les 'Essais optimistes.'" 4.
***Mais c'est Metchnikoff!*:** Unidentified Odessa newspaper, ARAN 584.1.334, 13.
in an amphitheater in the Sorbonne building: The Durkheim Amphitheater.
Metchnikoff Point: Geographic Names Information System, United States Board on Geographic Names, accessed May 13, 2015, http://geonames.usgs.gov/apex/f?p=gnispq:5:0::NO::P5_ANTAR_ID:9903.
each person is like a fraction: This statement appears in a book of aphorisms by writers and philosophers collected by Tolstoy (*Krug chteniia,* chapter "November 9") and apparently belongs to Tolstoy himself.
"Tubercular ganglions!": d'Hérelle, "Autobiographie," 168.
painters and sculptors asked Metchnikoff to pose: Here's a partial list: Two portraits, a 1907 one by Joseph Perelmann and a 1911 one by William Laparra, are at the Pasteur Institute, as is a statuette by Théodore Rivière. EM's portrait by Eugène Carrière is in the Pauls Stradins Museum for History of Medicine in Riga, Latvia, as are some of his portraits by Olga. The current location of EM's portrait by Leon Comerre is unknown; it was sold at an auction in June of 1906 (*Le Figaro,* June 9, 1906) for 1,950 francs (a painting of a woman's head by Renoir was sold for 2,550 francs at the same auction).
"I did my best to assume the pose": EM to OM, July 7, 1902. *Letters 2,* 101.
"The model had moved": Jean Pierrard, "Eugène Carrière, la vie en beige," *Le Point,* no. 1261, November 16, 1996, 121.
vision must have been impaired by kidney trouble: "Mechnikov kak myslitel'," unidentified Russian newspaper clipping, ARAN 584.2.215, 2.
artistic scene burgeoned right around the corner: For a vivid description of Montparnasse in the early twentieth century, see Bougault, *Paris Montparnasse.*
one of Metchnikoff's favorite spots in Paris: EM to OM, April 19, 1904. *Letters 2,* 137.
the only jungle he had ever seen: Bougault, *Paris Montparnasse,* 21.

La Ruche, "the hive": Ibid, 41.

The French continued to be infatuated with all things Russian: Foshko, *France's Russian Moment,* 71–83.

"The Slav loves to despair": Levaditi, "Élie Metchnikoff – Émile Roux," 7.

"Some of the more conservative temperament": "The Glycobacter," *New York Times,* June 18, 1912, 10.

Chapter 37: Triumphant Tour

a letter from Stockholm: Karl Mörner to EM, October 30, 1908. ARAN 584.2.148A.

"waiting list" for the Nobel: This concept was formulated by Luttenberger, "Arrhenius vs. Ehrlich," 144–5.

"He has the great merit": Lagerkvist, *Pioneers of Microbiology,* 164.

honored jointly with Paul Ehrlich: For a discussion of the Prize, including the obstacles put up by Ehrlich's enemies, see Lagerkvist, *Pioneers of Microbiology,* 154–68; Luttenberger, "Arrhenius vs. Ehrlich"; Silverstein, *Paul Ehrlich's Receptor Immunology,* 69–70; and Tauber, "The Birth of Immunology, III."

"Elie Metchnikoff was the first to take up": Karl Mörner, "Award Ceremony Speech." *Nobelprize.org,* Nobel Media AB 2014, accessed January 17, 2015, www.nobelprize.org/nobel_prizes/medicine/laureates/1908/press.html.

A more likely reason was the vexing pairing: I'm grateful to Arthur Silverstein for this insight (e-mail to author, March 2, 2014).

he believed he deserved a Nobel Prize all on his own: Luttenberger, "Arrhenius vs. Ehrlich," 164–65.

Ehrlich had received sixty nominations: Nomination Database. *Nobelprize.org,* accessed July 17, 2015, www.nobelprize.org/nomination/archive/.

Metchnikoff had accumulated sixty-six: France (21), Lvov, Prague and Krakow, then in Austria-Hungary (21), Russia, then including Odessa and Warsaw (10), Belgium (8), the United States (2), Switzerland (2), Italy (1) and Germany (1). Nomination Database. *Nobelprize.org,* accessed July 17, 2015, www.nobelprize.org/nomination/archive/.

"Metchnikoff was acknowledged as 'the founder'": Tauber, "The Birth of Immunology, III," 525.

"seen as guaranteeing the correctness of their respective theories": Lagerkvist, *Pioneers of Microbiology,* 164.

"financially, the Nobel Prize is of course much more": EM to Nikolai Chistovich, December 19, 1908. EM, *Pis'ma,* 206.

the purchase of apes: According to some reports (de Clap'ie, "Professor Il'ia Mechnikov," 64), Metchnikoff exclaimed upon receiving the news from Stokholm: "Now we'll have primates!"

"The enterprise bore no fruit": *ZIIM,* 184.

"They say here that the Swedish climate": EM to Marie and Émile Rémy, May 16, 1909. Saada Collection.

the Swedes know German better: Ibid.

"together with my excellent friend": EM, "Sur l'état actuel," 1.

"The phagocyte theory, founded": Ibid., 23.

"Yesterday it even snowed": EM to Marie and Émile Rémy, May 21, 1909. Saada Collection.

"I'm very shy before such a large crowd": Ibid.

"We all in Russia have been feeling gloomy": I. P. Pavlov, "Rech' na obiedinennom zasedanii vsekh meditsinskikh i biologicheskikh nauchnykh obshchestv goroda S.-Peterburga," *Polnoe sobranie sochinenii,* vol. 6 (Moscow: Akademiia Nauk SSSR, 1952), 313.

"None of us can protect ourselves": N. Dimov, "Prorok v svoem otechestve," *Birzhevye Vedomosti,* no. 11107, May 15, 1909, 2.

Chapter 38: Two Monarchs

Main sources for this chapter: EM, "Den' u Tolstogo"; *ZIIM,* 161–65; Goldenveizer, *Vblizi Tolstogo,* 269–72 and 348; Reznik, *Mechnikov,* 154–58; "Pis'mo Ol'gi Nikolaevny Mechnikovoi Vere Aleksandrovne Chistovich," *Nauka i Zhizn,* no. 1, 1967, 153–56.

"When it was learned that the great student": Bernstein, "Metchnikoff—'The Apostle.'"

death of stomach cancer at forty-five: Ivan Ilyich Metchnikoff died in 1881 (*SV,* 239).

probably a composite portrait: According to Semyon Reznik (interview with author, March 1, 2009), the story at least partly reflects the family life of Tolstoy's sister-in-law Tatiana Kuzminskaia, as can be deduced from her memoir, *Moia zhizn' doma i v Iasnoi Poliane* (Tula: Tul'skoe Knizhnoe Izdatel'stvo, 1958).

"For instance, as a professor of zoology": *SV,* 129.

"He even claimed to have desired death": OM to Behring, undated. AIP, MTO.1.

Chapter 39: A Metchnikoff Cow

"A surgical monstrosity": Shaw to T.B. Layton, March 13, 1948, quoted in Dally, *Fantasy Surgery,* 153.

the dexterity of a born mechanic: "Sir William Arbuthnot Lane (1856–1943)," The Historic Hospital Admission Records Project, accessed April 9, 2014, www.hharp.org/library/gosh/doctors/william-arbuthnot-lane.html.

Lane decided that the entire colon was a "vestige": T. B. Layton, *Sir William Arbuthnot Lane* (London: Livingstone, 1956), 89, cited from Podolsky, "Cultural Divergence," 8.

"the habit of keeping the trunk": W. Arbuthnot Lane, "Civilisation in Relation to the Abdominal Viscera, with Remarks on the Corset," *Lancet* 174, no. 4498 (1909): 1416.

a biographer of his called a "Great Debate": Layton, *Lane,* 99, cited from Podolsky, "Cultural Divergence," 8.

"It is hard to believe that a great structure": Arthur Keith, untitled remarks, "A Discussion on Alimentary Toxaemia; Its Sources, Consequences, and Treatment," *Proceedings of the Royal Society of Medicine* 6, Gen. Rep. (1913): 193.

"could not be discussed seriously": EM, "La Vieillesse," repr., *ASS,* vol. 15, 52.

"Dr. Lane, an unusually skillful and courageous": EM, "Mirosozertsanie i meditsina," *Vestnik Evropy* 1 (1910): 217–35, repr., *ASS,* vol. 13, 204.

"only serve to bring the whole subject": Wiley, *Foods and Their Adulteration,* 556.

"Professor Metchnikoff, of the Paris Pasteur Institute, is distinctly": "Our London Letter," *Medical News* 85, no. 3 (July 16, 1904): 132.

"Contrary to what many journalists have made me say": EM, "The Utility of Lactic Microbes," 58.

Professor Metchnikoff Discusses: "Professor Metchnikoff Discusses Youth at 100 Years: New Diet Experiments of the Pasteur Institute to Defer Old Age and Give Man an Enormously Prolonged Period of Individuality and Usefulness," *New York American,* June 30, 1912. ARAN 584.1.336.

Brush the Cow's Teeth: *Sunday Record-Herald,* Chicago, April 18, 1909. ARAN 584.1.336.

the *Daily Chronicle* mockingly cited "evidence": "Literary Notes," *British Medical Journal* 2, no. 2344 (1905): 1467.

"Metchnikoff cow": "Christmas Entertainments," *Times* (London), December 22, 1910, 11.

"I'm sixty-four and you see": Unidentified Russian newspaper, May 30, 1909. ARAN 584.2.181, 25-over.

"I've started eating Bulgarian sour milk": In "Les microbes lactiques," Metchnikoff provided more details: "To every meal, I add one or two pots of sour milk prepared with lactic bacilli, as well as dough from Bulgarian bacilli, which I eat with jam. As often as I can, I also eat dates, stuffed with Bulgarian bacilli or simply cleansed with boiling water." (repr., *ASS,* vol. 15, 274).

The bats did just fine when fed sterile food: Metchnikoff, Weinberg, Pozerski, Distaso, and Berthelot, "Roussettes et microbes," *Annales* 23, no. 12 (1909): 937–78.

tadpoles grown in sterile conditions: EM, "Les microbes intestinaux," *Bulletin de l'Institut Pasteur* 1, no. 6 (1903): 217–28 and no. 7 (1903): 265–82, repr., *ASS,* vol. 15, 108.

regularly examined his own urine: *ZIIM,* 166. ARAN has notebooks with EM's self-observations of his intestinal flora (584.1.217) and heart activity and urine as affected by diet (584.1.218 and 584.1.219).

response to food was indeed altered by the gut microbes: EM, "Vorzeitiges Altern und Stoffwechsel," *Neue Freie Presse,* no. 1785, May 31, 1914, 97–100, repr., *ASS,* vol. 15, 73–74.

"The idea is to define precisely the two categories": Ibid., 72.

richer in adults than in babies: EM, "Les microbes intestinaux," *ASS,* vol. 15, 99.

an Argentinian study in which calves: EM, "Études sur la flore intestinale. Deuxième mémoire. Poisons intestinaux et scléroses," *Annales* 24, no. 10 (1910): 766.

"immediate hygienic measures": EM, "Études sur la flore intestinale. Quatrième mémoire. Les diarrhées des nourrissons," *Annales* 28, no. 2 (1914): 119.

launched intensive investigations into lactic acid microbes: Rettger and Cheplin, *A Treatise on the Transformation.*

"futile fantasy": EM, "Introduction to the 4th Edition," *Etiudy o prirode,* 23.

"*Fouille-Merde,*" "Shit-Digger": d'Hérelle, "Autobiographie," 170.

"the ten greatest men now alive": "Who Are the Ten," *Strand Magazine.*

"The list on the whole is not so foolish": "The Ten Greatest Men Living," *British Medical Journal* 2, no. 2657 (1911): 1490.

Chapter 40: Rational Worldview

"a widespread, chauvinistic feeling": Weber, *The Nationalist Revival*, 95.

"Foreigners were accused": *ZIIM*, 182.

"If the profits from these sales": P.-L. Lafage, "À l'Institut Pasteur," *Le* XIXe *Siècle*, June 18, 1912.

"For a few years, under the influence of Metchnikoff": "Revue des sciences," *Journal des Débats Politiques et Littéraires*, May 30, 1912, 1.

"on the lunatic fringe": Weber, *The Nationalist Revival*, 100.

a microbe he called *glycobacter*: EM and Eugène Wollman, "Sur quelques essais de désintoxication intestinale," *Annales* 26, no. 11 (1912): 843.

"the Jew of dog droppings": "Échos: Les merveilleux remèdes d'Elie Metchnicrotte," *L'Action Française*, July 1, 1912, 1. See also Leon Daudet, "Elie Metchnikoff, un juif de crottes de chien," *L'Action Française*, June 21, 1912, 1.

"for my friend Metchnikoff, who for twenty-four years": Roux, letter to the editor, "Correspondance," *Journal des Débats*, July 4, 1912, 3.

"With time, when science eliminates misery": EM, *Sorok let*, 40.

Chapter 41: The Last War

"Fields with an endless horizon": *ZIIM*, 184.

"depends on the organism itself, its most inner essence": EM, *Etiudy o prirode*, 257.

Metchnikoff's last published study: EM, "La mort du papillon du mûrier: Un chapitre de thanatologie," *Annales* 29, no. 10 (1915): 477–96.

"Appeals for peace are now heard": EM, "Obzor glavneishikh uspekhov nauki o mikrobakh za 1909 god," *Russkie Vedomosti*, January 1, 1910, 13.

"This episode marked him for life": *ZIIM*, 13.

"I'll never forget his return home": *ZIIM*, 196.

the entire Pasteur Institute was recruited: Calmette, "L'oeuvre de l'Institut Pasteur," 463–82.

the Pasteur Institute supplied 670,000 doses: Ibid., 466.

twenty thousand ampules per day, which gives an idea: Gascar, *Du côté de chez Monsieur*, 204.

"All that's left in Paris": Jean Ajalbert, *Dans Paris, la grande ville (sensations de guerre)* (Paris: Éditions Georges Crès et cie, 1916), 9, quoted in Quinn, *Marie Curie*, 358.

"Let us hope that this unparalleled carnage": EM, *Osnovateli*, 159.

"This war will certainly": EM to Elie Rémy, September 2, 1914. Saada Collection.

"The war cut him down like an oak": OM's diary, May 16, 1944. ARAN 584.6.4, 34-over.

started calling him Père Noël: *ZIIM*, 199.

Chapter 42: A True Benefactor

"One shouldn't forget that I started living": EM, "Dnevnik s zapisiami," October 19, 1913, p. 4.

"Let all those who expected me": Ibid., 5.

admitted to a visitor from Russia: Antsiferov, *Iz dum o bylom,* 271.
"Now I'm positively unafraid of death": EM, "Dnevnik s zapisiami," May 16, 1915, p. 11.
"Throughout his illness, his warmth": *ZIIM,* 209.
"My dream to die quickly": EM, "Dnevnik s zapisiami," May 16, 1916, p. 13.
"can serve as proof that my orthobiosis": Ibid., June 18, 1916, p. 17.
"*Donc alors,* this is my last week": d'Hérelle, "Autobiographie," 171.
On the afternoon of July 15: In her book, OM doesn't provide an exact date for EM's death, and certain sources mistakenly state it as July 16. According to his death certificate, issued by the Mairie du Quinzième Arrondissement de Paris, EM died at 5:30 PM on July 15, 1916 (ARAN 584.2.208).
"You are a friend": *ZIIM,* 218.
"I looked up": Ibid.
Lili later recalled that throughout the hour-long: Saada-Rémy, "Souvenirs de Lili," 31.
"He was one of ours": Bianchon, "Elie Metchnikoff."
"one of the commanding old men": "Elie Metchnikoff," *Nation,* July 20, 1916, 53.
"one of the most remarkable figures": Lankester, "Elias Metchnikoff," 443.
"the whole scientific world will mourn": "Obituary," *British Medical Journal,* 130.
"However untenable some of the conclusions": "Elie Metchnikoff," *Nation,* July 20, 1916, 54.
"He harped on the theme so much": "The Man Who Tried to Lengthen Life," *Literary Digest,* New York, October 7, 1916. ARAN 584.2.216, 4.
"The death of Professor Metchnikoff, for whom": "Metchnikoff and Old Age," *Sunday Times,* July 23, 1916, ARAN 584.2.216, 24.
"These cranky ideas": *Medical Times,* July 22, 1916, ARAN 584.2.216, 44.
"as a discoverer and a true benefactor": "Metchnikoff," *New York Times,* July 17, 1916, 10.
"It would be interesting to compile a list": "The Dean of Bacteriologists," *Scientific American,* 98.

Chapter 43: Olga's Crime

"I don't know if I actually cried": OM's diary, ARAN 584.6.5, 8-over.
a total of 30,000 francs: Cressac, *Le Docteur Roux,* 171. The catalogue of the 1923 exhibition is in ARAN 584.6.41.
His involvement with Olga . . . came only second: Cressac, *Le Docteur Roux,* 107.
"At least in appearance, Metchnikoff's passing": Ibid., 171.
her husband's museum: The former Metchnikoff Moscow Bacteriology Institute, now the Metchnikoff Biomed company, still has a Metchnikoff museum, but most exhibits are copies. The original documents are in the Archive of the Russian Academy of Sciences (ARAN); the rest of the collection is in the Pauls Stradins Museum for History of Medicine in Riga, Latvia.
"All the portraits are assembled in the museum": OM to Lili, October 20, 1935. Saada Collection.
She had asked to be cremated: Draft of OM's letter to her brother containing her will, ARAN 584.6.5, 7.
"There's room there," she explained: Ibid.
managed to have them retrieved from the south of France and transferred: Gaisinovich, "Trudy i dni," 11–14.

Chapter 44: Vanishing

"Every morning I watched a boy": Alexandra Tolstaia, *Daughter* (Moscow: Vagrius, 2000), 462.

"We used tinned-copper vats to heat the milk": *The Danone Story: 1919–2009* (Danone Corporate Communications, 2009), 5.

"Danone yogurt. Prepared as recommended": Ibid., p. 6.

"Thousands of persons felt a keen sense": Kellogg, *Autointoxication*, 307–13.

sulfa drugs, and then antibiotics, pushed sour milk: Podolsky, "The Art of Medicine," 1811.

they also killed the interest in the gut flora: Lee and Brey, "How Microbiomes Influence," 7.1.

"It would seem that in his views": Petrie, "The Scientific Work," 548.

immunology had run out of easy successes: Arthur M. Silverstein, "Autoimmunity *Versus Horror Autotoxicus*: The Struggle for Recognition," *Nature Immunology* 2, no. 4 (2001): 280.

"Why did you break Professor Metchnikoff?": Bulgakov, *Sobach'e serdtse*, chapter 4.

"Metchnikoff's works have brought glory": Miterev, "Il'ia Il'ich."

"I. I. Metchnikoff's works serve as a constant reference": L. A. Zil'ber, "Immunologicheskie issledovaniia I. I. Mechnikova," in *ASS*, vol. 7, 544.

"attempts by certain bourgeois scientists": Milenushkin, "I. I. Mechnikov i ego," 536.

"an immunological paradox": Janeway et al., "Hypergammaglobulinemia," 388.

In the words of one prominent researcher: Niels Jerne (Michael Sela, interview with author, November 4, 2008).

the book of the conference's proceedings: Karnovsky and Bolis, *Phagocytosis—Past and Future.*

"The antibodies and later the T cells clearly captivated": Ralph Steinman, interview with author, December 19, 2008.

"In the 1980s, the majority of smart people": Ira Mellman, interview with author, July 25, 2011.

nobody worked on "Metchnikoff's" innate immunity: Jean-Marc Cavaillon, interview with author, November 8, 2012.

PART V: LEGACY

Chapter 45: Metchnikoff's Life

she converted to Judaism: Lili was given the Hebrew name Miriam, which apparently she never used.

Chapter 46: Metchnikoff's Policemen

recent tributes to Metchnikoff, marking the centennial: See, for example, Kaufmann, "Immunology's Foundation."

report on the festive two-day symposium: Nathan, "Metchnikoff's Legacy."

cover of one of the issues of the *Journal*: Cavaillon, "The Historical Milestones."

"Metchnikoff made the first essential step": Jules Hoffmann, interview with author, June 23, 2014.

"rediscovery" of the innate component: See, for example, Germain, "An Innately Interesting Decade," 1316.

some twenty-five thousand scientific papers have centered on TLRs: Hoffmann, interview with author.

Hoffmann began the historical overview: Jules A. Hoffmann, Nobel Lecture, "The Host Defense of Insects: A Paradigm for Innate Immunity," *Nobelprize.org*, Nobel Media AB 2014, accessed September 11, 2015, www.nobelprize.org/nobel_prizes/medicine/laureates/2011/hoffmann-lecture.html.

"It's really silly—as Metchnikoff and Ehrlich": Ralph Steinman, interview with author, December 19, 2008.

***Salmonella* germs are gobbled up:** J. Farache et al., "Luminal Bacteria Recruit CD103(+) Dendritic Cells into the Intestinal Epithelium to Sample Bacterial Antigens for Presentation," *Immunity* 38, no. 3 (2013): 581–95.

he was often amazingly prescient: Varol, Mildner, and Jung, "Macrophages: Development and Tissue Specialization."

aptly titled "Metchnikoff's Policemen": Stefater et al., "Metchnikoff's Policemen."

the most abundant single cell type: David A. Hume, "*The Biology of Macrophages—An Online Review,*" Edition 1.1, May 2012, accessed May 27, 2015, www.macrophages.com/macrophage-review.

macrophages, which make up about one-fifth of all liver cells: Siamon Gordon, e-mail to author, June 28, 2015.

"Macrophages are involved in almost every disease": Wynn, Chawla, and Pollard, "Macrophage Biology," 452.

"corrupt" the macrophages: Bonavita et al., "Phagocytes as Corrupted Policemen."

drugs aimed at blocking these "corrupted policemen": Eduardo Bonavita et al., "PTX3 Is an Extrinsic Oncosuppressor Regulating Complement-Dependent Inflammation in Cancer," *Cell* 160 (2015): 700–14.

Chapter 47: Ultimate Closure

Dina's story is based on an interview with author, March 20, 2015.

defined as a syndrome called "fatal granulomatous": Tracy Assari, "Chronic Granulomatous Disease; Fundamental Stages in Our Understanding of CGD," *Medical Immunology* 5:4 (2006), accessed September 11, 2015, doi:10.1186/1476-9433-5-4.

"Metchnikoff talked about phagocytosis": Baruch Wolach, interview with author, March 18, 2015.

immune deficiency varies widely: See, for example: Joseph Ben-Ari, Ofir Wolach, Ronit Gavrieli, and Baruch Wolach, "Infections Associated with Chronic Granulomatous

Disease: Linking Genetics to Phenotypic Expression," *Expert Review of Anti-Infective Therapy* 10, no. 8 (2012): 881–94.

one of the leading gene therapy trials for CGD: Marion G. Ott et al., "Correction of X-linked Chronic Granulomatous Disease by Gene Therapy," *Nature Medicine* 12, no. 4 (2006): 401–9, accessed September 11, 2015, doi:10.1038/nm1393; Manuel Grez, interview with author, February 13, 2013.

Chapter 48: Living to 150

"What is the maximum age": EM, *Etiudy optimizma,* 86.

the longevity record for animals: Silvertown, *The Long and the Short,* Kindle edition, loc. 2165.

"We do agree on one thing": Steven Austad, interview with author, March 30, 2015.

"inflamm-aging": See, for example, Claudio Franceschi and Judith Campisi, "Chronic inflammation (inflammaging) and its potential contribution to age-associated diseases," *Journals of Gerontology. Series A: Biological Sciences* 69 (June 2014): Suppl 1:S4-9, doi: 10.1093/gerona/glu057.

through studies of invertebrates—worms and flies: Heintz and Mair, "You Are What You Host."

"The intestinal microbiome has become": S. Jay Olshansky, interview with author, March 20, 2015.

"attempts at 'probiotics' seemed a marginal practice": Podolsky, "The Art of Medicine," 1810.

"Probiotics: 100 Years (1907–2007) After Elie Metchnikoff's Observation": Kingsley C. Anukam and Gregor Reid, in *Communicating Current Research and Educational Topics and Trends in Applied Microbiology,* vol. 1, ed. A. Méndez-Vilas (Badajorz: Formatex, 2007), 466–74.

"a mythical character": Gregor Reid, interview with author, April 27, 2015.

"the grandfather of modern probiotics research": Simin Meydani, interview with author, March 18, 2015.

In 2002, a paper in *Science* announced: Oeppen and Vaupel, "Broken Limits," 1029. See also Vaupel, "Biodemography."

the bulk of the reduction in mortality had occurred since 1900: Burger, Baudisch, and Vaupel, "Human Mortality Improvement," 18210.

has been remarkable: about thirty years: Christensen et al., "Ageing Populations," 1196.

In the United States, for example, between 1900 and 2010: Arias, "United States Life Tables," 45–46.

celebrate their one hundredth birthday: Christensen et al., "Ageing Populations," 1196.

Future gains in life span are likely to grow progressively smaller: For an explanation, including an analogy with the history of the one-mile run, see Olshansky and Carnes, *A Measured Breath of Life.*

"It is difficult to imagine": EM, *Sorok let,* 21.

"tried to force his beliefs down nature's throat": de Kruif, *Microbe Hunters,* 203.

BIBLIOGRAPHY

Ackerknecht, Erwin H. *Rudolf Virchow: Doctor, Statesman, Anthropologist.* Madison: University of Wisconsin Press, 1953.

Ambrose, Charles T. "The Osler Slide, a Demonstration of Phagocytosis from 1876: Reports of Phagocytosis Before Metchnikoff's 1880 Paper." *Cellular Immunology* 240 (2006): 1–4.

Aminoff, Michael J. *Brown-Séquard: An Improbable Genius Who Transformed Medicine.* New York: Oxford University Press, 2010.

Antonovich, M. A., and G. Z. Eliseev. *Shestidesiatye gody.* Moscow: Akademia, 1933.

Antsiferov, N. P. *Iz dum o bylom: Vospominaniia.* Moscow: Feniks, 1992.

Arias, Elizabeth. "United States Life Tables, 2010." *National Vital Statistics Reports* 63, no. 7 (2014): 1–54. Accessed September 11, 2015. www.cdc.gov/nchs/data/nvsr/nvsr63/nvsr63_07.pdf.

Bainville, Jacques. *Histoire de deux peuples continuée jusqu'à Hitler.* Paris: Artheme Fayard, 1933.

Bardakh, Y. Y. "Il'ia Il'ich Mechnikov." *Vrachebnoe Delo* 15–17 (1925): 1195–201.

Basinskii, Pavel. *Lev Tolstoi: Begstvo iz raia.* Moscow: AST Astrel', 2011.

Basset, Serge. "Vive la Vie!" *Le Matin,* December 27, 1899, 1.

Baumgarten, Paul. "Referate." *Berliner klinische Wochenschrift* (December 15, 1884): 802–4; (December 22, 1884): 818–19.

Baumgarten, Paul. "Zur Kritik der Metschnikoff'schen Phagocytentheorie." *Zeitschrift für klinische Medicin* 15 (1889): 1–41.

Baumgarten, Paul. "Über das 'Experimentum crusis' der Phagocytenlehre." *Beiträge zur pathologischen Anatomie* 7 (1890): 1–10.

Bäumler, Ernst. *Paul Ehrlich: Scientist for Life.* Translated by Grant Edwards. New York: Holmes & Meier, 1984.

Baxter, Alan G. "Louis Pasteur's Beer of Revenge." *Nature Reviews—Immunology* 1 (December 2001): 229–32.

Behring, Emil. "On the Cause of the Immunity of Rats Against Anthrax." *Centralblatt für klinische Medicin* (September 22, 1888): 681–90. Translated in Linton, *Emil von Behring,* 442–49.

Behring, Emil, and Shibasaburo Kitasato. "Über das Zustandekommen der Diphtherie-Immunität und der Tetanus-Immunität bei Thieren." *Deutsche Medizinische Wochenschrift,* December 4, 1890. Translated in Linton, *Emil von Behring,* 450–44.

Bénézit, E. *Dictionnaire critique et documentaire des peintres, sculpteurs, dessinateurs et graveurs,* vol. 6. Paris: Librairie Gründ, 1966, s.v. "Metchnikoff (Mme Olga)."

Benjamin, Walter. "Paris, Capital of the Nineteenth Century." In *Essays, Aphorisms, Autobiographical Writings,* 146–62. New York: Schocken Books, 1986.

Berkhin, I. B., and I. A. Fedosov. *Istoriia SSSR: Uchebnoe posobie dlia 9-go klassa.* Moscow: Prosveshchenie, 1974.

Bernstein, Herman. "Metchnikoff—'The Apostle of Optimism'—On the Science of Living." *New York Times*, August 1, 1909, SM4.

Besredka, Alexandre. "Vospominaniia ob I. I. Mechnikove." *Priroda* 30 (1926): 37–42.

Beumer, J. "Jules Bordet 1870–1961." *Journal of General Microbiology* 29, no. 1 (1962): 1–13.

Bial, Andrea K. "Optimism and Aging." *Journal of the American Geriatrics Society* 53, no. 3 (2005): 541–42.

Bianchon, Horace. "Les 'Essais optimistes' de M. Metchnikoff." *Le Figaro*, April 19, 1907, 4.

Bianchon, Horace. "Elie Metchnikoff." *Le Figaro*, July 16, 1916, 1.

Binoche, Jacques. *Histoire des relations franco-allemandes de 1870 à nos jours*. Paris: Masson, 1996.

Bogdanovich, T. A. *Liubov' liudei shestidesiatykh godov*. Leningrad: Academia, 1929.

Bonavita, Eduardo, Maria Rosaria Galdiero, Sebastien Jaillon, and Alberto Mantovani. "Phagocytes as Corrupted Policemen in Cancer-Related Inflammation." *Advances in Cancer Research* 128 (2015): 141–71.

Bougault, Valerie. *Paris Montparnasse: The Heyday of Modern Art, 1910–1940*. Paris: Finest SA, Editions Pierre Terrail, 1997.

Brandreth, Charles J. "The Man Who Prolongs Life." *London Magazine*, 23, no. 137 (January 1910), 583.

Brieger, Gert. "Introduction," ix–xxxi. In Elie Metchnikoff, *Immunity in Infective Diseases*. Translated from the French by Francis G. Binnie. New York: Johnson Reprint Corporation, 1968.

British Medical Journal. "Obituary. Elias Metchnikoff." Vol. 2, no. 2899 (1916): 130–31.

Brock, Thomas D. *Robert Koch: A Life in Medicine and Bacteriology*. Madison, WI: Science Tech Publishers, 1988.

Bulgakov, M. A. *Sobach'e serdtse: Chudovishchnaia istoriia*. 1925. Accessed September 9, 2015. www.vehi.net/mbulgakov/sobach.html.

Burger, Oskar, Annette Baudisch, and James W. Vaupel. "Human Mortality Improvement in Evolutionary Context." *Proceedings of the National Academy of Sciences, USA* 109, no. 44 (2012): 18210–4.

Burnet, Etienne. *Un Européen Elie Metchnikoff*. Tunis: Editions Calypso, 1966.

Calmette, Albert. "L'oeuvre de l'Institut Pasteur pendant la guerre." *Revue d'hygiène et de police sanitaire* 43 (1921): 463–82.

Cambrosio, Alberto, Daniel Jacobi, and Peter Keating. "Ehrlich's 'Beautiful Pictures' and the Controversial Beginnings of Immunological Imagery." *Isis* 82 (1993): 662–99.

Cavaillon, Jean-Marc. "The Historical Milestones in the Understanding of Leukocyte Biology Initiated by Elie Metchnikoff." *Journal of Leukocyte Biology* 90 (2011): 413–24.

Chernyak, Leon, and Alfred I. Tauber. "The Birth of Immunology: Metchnikoff, the Embryologist." *Cellular Immunology* 117:1 (1988): 218–33.

Chernyshevsky, Nikolai. *What Is to Be Done?* Translated by Michael R. Katz. Ithaca, NY: Cornell University Press, 1989.

Chistovich, F. Ia. "Pamiati L. A. Tarasevicha." *Zhurnal Mikrobiologii, Patologii i Infektsionnykh Boleznei* 4, no. 4 (1927): 343–51.

Chistovich, N. Ia. "Iz zapisok N. Ia. Chistovicha." ARAN 584.2.30, 6 p.

Chistovich, N. Ia. *I. I. Mechnikov*. Berlin: RSFSR Gosudarstvennoe Izdatel'stvo, 1923.

Christensen, Kaare, Gabriele Doblhammer, Roland Rau, and James W. Vaupel. "Ageing Populations: The Challenges Ahead." *Lancet* 374, no. 9696 (2009): 1196–208.

Cochrane, Al. "Elie Metchnikoff and His Theory of an 'Instinct de la Mort.'" *International Journal of Psychoanalysis* 15 (1934): 265–70. Reprinted in *International Journal of Epidemiology* 32 (2003): 32–34.

Cope, Zachary. *The Versatile Victorian: Being the Life of Sir Henry Thompson, bt. 1820–1904.* London: Harvey & Blythe Ltd., 1951.

Cressac, Mary. *Le Docteur Roux, Mon Oncle.* Paris: L'Arche, 1950.

Crist, Eileen, and Alfred I. Tauber. "Debating Humoral Immunity and Epistemology: The Rivalry of the Immunochemists Jules Bordet and Paul Ehrlich." *Journal of the History of Biology* 30 (1997): 321–56.

Dally, Ann. *Fantasy Surgery, 1880–1930: With Special Reference to Sir William Arbuthnot Lane.* Amsterdam: Rodopi, 1996.

de Clap'ie, Olga. "Professor Il'ia Mechnikov." *Vozrozhdenie* (Paris), no. 190, October 1967, 53–69.

de Kruif, Paul. *Microbe Hunters.* New York: Harcourt Brace & Co., 1926, reprinted in 1996.

Delaunay, Albert. *L'Institut Pasteur des origines à aujourd'hui.* Paris: Éditions France-Empire, 1962.

Delaunay, Albert. "A propos de la famille Saada." December 1969. AIP, DEL.D2, 8 p.

Delaunay, Albert. "Jules Bordet et l'Institut Pasteur de Paris." *Histoire de la Medicine,* April 1971, 2–45.

d'Hérelle, Félix. "Autobiographie." AIP, HER.1, 515 p.

Douglas, Loudon. *The Bacillus of Long Life.* New York: G. P. Putnam's Sons, 1911. Accessed August 30, 2015. www.gutenberg.org/ebooks/31691.

Dubos, René, and Jean Dubos. *The White Plague: Tuberculosis, Man, and Society.* New Brunswick, NJ: Rutgers University Press, 1996.

Duclaux, Émile. *Pasteur: The History of a Mind.* Translated by Erwin F. Smith and Florence Hedges. Philadelphia: W. B. Saunders Company, 1920.

Dunnill, Michael. *The Plato of Praed Street.* London: Royal Society of Medicine Press, 2000.

Ehrlich, Paul. "Croonian Lecture: On Immunity with Special Reference to Cell Life." *Proceedings of the Royal Society of London* 66 (1899–1900): 424–48.

Ehrlich, Paul. *Studies on Immunity.* Translated by Charles Bolduan. New York: John Wiley & Sons, 1910.

Eksteins, Modris. *Rites of Spring.* New York: Doubleday, 1989.

Fedosov, I. A., ed. *Istoriia SSSR (XIX nachalo XX vekov).* Moscow: Vysshaia Shkola, 1981.

Fisher, David C. "The Rise of the Radical Intelligentsia, 1862–1881." In *Events That Changed Russia Since 1855,* edited by Frank W. Thackeray, 23–43. Westport, CT: Greenwood Press, 2007.

Foshko, Katherine. *France's Russian Moment.* PhD diss., Yale University, 2008.

Frith, John. "Syphilis—Its Early History and Treatment Until Penicillin and the Debate on Its Origins." *Journal of Military and Veterans' Health* 20, no. 4 (November 2012). Accessed August 31, 2015. http://jmvh.org/issue/volume-20-no-4/.

Gaillard, Marc. *Paris: Les Expositions Universelles de 1855 à 1937.* Paris: Les Presses Franciliennes, 2005.

Gaisinovich, A. E. "Trudy i dni I. I. Mechnikova," 3–42. In I. I. Mechnikov. *Pis'ma k Ol'ge Mechnikovoi, 1900–1914.* Edited by A. E. Gaisinovich and B. V. Levshin. Moscow: Nauka, 1980.

Gaisinovich, A. E., and A. K. Panfilova. "Epistoliarnoe nasledie I. I. Mechnikova," 3–38. In I. I. Mechnikov. *Pis'ma (1863–1916).* Moscow: Nauka, 1974.

Gascar, Pierre. *Du côté de chez Monsieur Pasteur.* Paris: Éditions Odile Jacob, 1986.

Germain, Ronald N. "An Innately Interesting Decade of Research in Immunology." *Nature Medicine* 10, no. 12 (2004): 1307–20.

Gessen, Yulii. *Evrei v Rossii.* St. Petersburg, 1906.

Ghiselin, Michael T., and Christiane Groeben. "Elias Metschnikoff, Anton Dohrn, and the Metazoan Common Ancestor." *Journal of the History of Biology* 30 (1997): 211–28.

Godlee, Sir Rickman John. *Lord Lister.* London: MacMillan and Co., 1918.

Goetz, Thomas. *The Remedy: Robert Koch, Arthur Conan Doyle, and the Quest to Cure Tuberculosis.* New York: Gotham, 2014.

Goldenveizer, A. B. *Vblizi Tolstogo.* Moscow: Gosudarstvennoe Izdatel'stvo Khudozhestvennoi Literatury, 1959.

Gordin, Michael D. *A Well-Ordered Thing: Dmitrii Mendeleev and the Shadow of the Periodic Table.* New York: Basic Books, 2004.

Gordon, Siamon. "Elie Metchnikoff: Father of Natural Immunity." *European Journal of Immunology* 38 (2008): 3257–64.

Grmek, Mirko Dražen. *On Ageing and Old Age, Basic Problems and Historic Aspects of Gerontology and Geriatrics.* Monographiae Biologicae, vol. 5, no. 2. The Hague: Uitgeverij Dr. W. Junk, 1958.

Gourko, Helena, Donald L. Williamson, and Alfred Tauber. *The Evolutionary Biology Papers of Elie Metchnikoff.* Dordrecht: Kluwer Academic Publishers, 2000.

Groeben, Christiane, and Klaus Wenig, eds. *Anton Dohrn und Rudolf Virchow: Briefwechsel 1864–1902.* Berlin: Akademie Verlag, 1992.

Gruman, Gerald J. *A History of Ideas About the Prolongation of Life: The Evolution of Prolongevity Hypothesis to 1800.* Transactions of the American Philosophical Society, vol. 56, part 9. Philadelphia: American Philosophical Society, 1966.

Grundbacher, F. J. "Behring's Discovery of Diphtheria and Tetanus Antitoxins." *Trends in Immunology* 13, no. 5 (1992): 188–90.

Gundobin, N. *Detskaia smertnost' v Rossii.* St. Petersburg, 1906.

Guriel, B. "Lenin and Weizmann: A Story of One Encounter." *Ha'aretz* (in Hebrew), November 3, 1967.

Gutnov, Dmitrij A. "L'École russe des hautes études sociales de Paris (1901–1906)." *Cahiers du monde russe* 43/2–3 (April–September 2002): 375–400. Accessed August 31, 2015. http://monderusse.revues.org/71.

Hage, Jerald, and Jonathon Mote. "Transformational Organizations and a Burst of Scientific Breakthroughs: The Institut Pasteur and Biomedicine, 1889–1919." *Social Science History* 34:1 (Spring 2010): 13–46.

Hansen, Bert. "America's First Medical Breakthrough." *The American Historical Review* 103, no. 2 (April 1998): 373–418.

Hastings, E. G., and B. W. Hammer. "The Occurrence and Distribution of a Lactic Acid Organism Resembling the *Bacillus Bulgaricus* of Yogurt," 197–206. In *Twenty-Fifth and Twenty-Sixth Annual Reports of the Agricultural Experiment Station of the University of Wisconsin, 1908–9.* Madison, WI: Democrat Printing Company, 1910.

Hausser, Elisabeth. *Paris au jour le jour, 1900–1919.* Paris: Les Éditions de Minuit, 1968.

Heintz, Caroline, and William Mair. "You Are What You Host: Microbiome Modulation of the Aging Process." *Cell* 156 (2014): 408–11.

Herschell, George. "On the Use of Selected Lactic Acid Bacilli and Soured Milk in the Treatment of Some Forms of Chronic Ill-Health." *Lancet* 172, no. 4432 (1908): 371–74.

Hippocrates. *Volume VII. Epidemics.* Edited and translated by Wesley D. Smith. Cambridge, MA: Harvard University Press, 1994.

Howard, Michael. *The Franco-Prussian War.* London: Collins, 1961.

Janeway, Charles A., John Craig, Murray Davidson, William Downey, David Gitlin, and Julia Sullivan. "Hypergammaglobulinemia Associated with Severe Recurrent and Chronic Nonspecific Infection." *American Journal of Diseases of Children* 88 (1954): 388–92.

Joanne, Adolphe. *Paris illustré: Nouveau guide de l'étranger et du Parisien.* Paris: Librairie de l. Hachette, 1867.

Jones, Colin. *Biography of a City.* New York: Penguin Books, 2004.

Johnson, Steven. *Where Good Ideas Come From.* New York: Riverhead Books, 2010.

Kandel, Felix. *Kniga vremen i sobytii,* vol. 1. Jerusalem: Tarbut, 1990.

Karnovsky, Manfred L., and Liana Bolis, eds. *Phagocytosis—Past and Future.* New York: Academic Press, 1982.

Kaufmann, Stefan H. E. "Immunology's Foundation: The 100-Year Anniversary of the Nobel Prize to Paul Ehrlich and Elie Metchnikoff." *Nature Immunology* 9, no. 7 (July 2008): 705–12.

Kellogg, John Harvey. *Autointoxication or Intestinal Toxemia.* Battle Creek, MI: Modern Medicine Publishing Co., 1919.

Khudozhniki russkogo zarubezh'ia, 1917–1939. A biographical dictionary. St. Petersburg: Notabene, 1999, s.v. "Mechnikova."

Koch, Robert. "An Address on Bacteriological Research." *British Medical Journal* 2, no. 1546 (1890): 380–83.

Kolodziejczyk, Ryszard. "Leon Newachowicz." *Rocznik Warszawski* 8 (1970): 143–73.

Kondakov, N. P. "Vospominaniia i dumy." ARAN 584.2.19, 11 p.

Kovalevskaia, Sof'ia. *Vospominaniia detstva i avtobiograficheskie ocherki.* Moscow: USSR Academy of Sciences, 1945.

Lagerkvist, Ulf. *Pioneers of Microbiology and the Nobel Prize.* Singapore: World Scientific Publishing, 2003.

Lagrange, Émile. *Monsieur Roux.* Brussels: A. D. Goemaere, 1954.

Lancet. "Obituary: Elie Metchnikoff." Vol. 188, no. 4847 (1916): 159–60.

Lankester, E. Ray. "Elias Metchnikoff." *Nature* 97, no. 2439 (1916): 443–46.

Lee, Won-Jae, and Paul T. Brey. "How Microbiomes Influence Metazoan Development: Insights from History and *Drosophila* Modeling of Gut-Microbe Interactions." *Annual Review of Cell and Developmental Biology* 29 (2013): 7.1-22.

Les Prix Nobel en 1908. Stockholm: Imprimerie royale, 1909.

Levaditi, Constantin. "Élie Metchnikoff–Émile Roux." Report at a congress in Paris dedicated to 30 years of the discovery of local prophylaxis of syphilis, March 12, 1936. ARAN 584.1.330, 8 p.

Levaditi, Constantin. "Centenaire d'Elie Metchnikoff." *La Presse Médicale.* July 30, 1945.

Lindenmann, Jean. "Origin of the Terms 'Antibody' and 'Antigen.'" *Scandinavian Journal of Immunology* 19 (1984): 281–85.

Linton, Derek S. *Emil von Behring: Infectious Disease, Immunology, Serum Therapy.* Philadelphia: American Philosophical Society, 2005.

Lister, Joseph. "An Address on the Present Position of Antiseptic Surgery." *British Medical Journal* 2, no. 1546 (1890): 377–79.

Lister, Joseph. "Lecture on Koch's Treatment of Tuberculosis." *Lancet* 136, no. 3511 (1890): 1257–59.

Lister, Joseph. "Presidential Address on the Relations of Clinical Medicine to Modern Scientific Development." *British Medical Journal* 2, no. 1864 (1896): 733–41.

Luttenberger, Franz. "Arrhenius vs. Ehrlich on Immunochemistry: Decisions About Scientific Progress in the Context of the Nobel Prize." *Theoretical Medicine* 13 (1992): 137–73.

Luttenberger, Franz. "Excellence and Chance: The Nobel Prize Case of E. von Behring and É. Roux." *History and Philosophy of the Life Sciences* 18 (1996): 225–38.

Mackenna, Stephen. "Dr. Metchnikoff." *Los Angeles Times*, February 18, 1900, IM6.

Maitron, Jean. *Le movement anarchiste en France*, vol. 1. Paris: François Maspero, 1975.

Markel, Howard. "The Medical Detectives." *New England Journal of Medicine* 353 no. 23 (2005): 2426–28.

Marquardt, Martha. *Paul Ehrlich*. London: William Heinemann Medical Books Ltd., 1949.

Mason, Frédéric S. "Notes on the Lactic Ferments as Therapeutic Agents." *Lancet* 172, no. 4439 (1908): 957–59.

Mayr, Ernst. *The Growth of Biological Thought*. Cambridge, MA: Belknap Press, 1982.

McIntyre, Neil. "The Marriage (1878) of Emile Roux (1853–1933) and Rose Anna Shedlock (b. c. 1850)." *Journal of Medical Biography* 16 (2008): 175–76.

Mechnikova, O. N. *Zhizn' Il'i Il'icha Mechnikova*. Moscow-Leningrad: Gosizdat, 1926.

Metchnikoff, Elie. "Ocherk voprosa o proiskhozhdenii vidov." *Vestnik Evropy*, 2(3): 68–134, 2(4): 715–47, 3(5): 117–49, 4(7): 158–97, 4(8): 567–606, 1876. Reprinted in *ASS*, vol. 4, 155-327.

Metchnikoff, Elie. "O tselebnykh silakh organizma." *Protokoly VII Siezda russkikh estestvoispytatelei i vrachei v Odesse, s 18 po 28 avg. 1883 g.* Odessa, 1883, suppl. 6, 1–14. Reprinted in *ASS*, vol. 13, 125–32.

Metchnikoff, Elie. "La théorie des phagocytes au Congrès hygiénique de Londres." *Annales de l'Institut Pasteur* 5, no. 8 (1891): 534–42.

Metchnikoff, Elie. "Zakon zhizni." *Vestnik Evropy* 5 (1891): 228–60. Reprinted in *ASS*, vol. 13, 133–58.

Metchnikoff, Elie. *Leçons sur la pathologie comparée de l'inflammation, faites à l'Institut Pasteur en avril et mai 1891*. Paris: Masson, 1892.

Metchnikoff, Elie. *Lectures on the Comparative Pathology of Inflammation*. Translated by F. A. Starling and E. H. Starling. London: Kegan Paul, Trench, Trübner & Co., 1893. Reprint, New York: Dover, 1968.

Metchnikoff, Elie. "Revue de quelques travaux sur la dégénérance senile." *Année Biologique* 3 (1897, published in 1899): 249–67.

Metchnikoff, Elie. "Sur la flore du corps humain (The Wilde Lecture)." *Memoirs and Proceedings of the Manchester Literary and Philosophical Society* 45, no. 5 (1901): 1–38.

Metchnikoff, Elie. *The Nature of Man: Studies in Optimistic Philosophy*. Translated from the French by P. Chalmers Mitchell. New York and London: G. P. Putnam's Sons, 1903.

Metchnikoff, Elie. *Nevospriimchivost' v infektsionnykh bolezniakh*. St. Petersburg: K. L. Rikker, 1903. Reprint, Moscow: Gosudarstvennoe Izdatel'stvo Meditsinskoi Literatury, 1947.

Metchnikoff, Elie. "La Vieillesse." *Revue Scientifique* 2, no. 3 (1904): 65–70 and no. 4 (1904): 100–105.

Metchnikoff, Elie. *Immunity in Infective Diseases*. Translated from the French by Francis G. Binnie. Cambridge: Cambridge University Press, 1905.

Metchnikoff, Elie. *Quelques Remarques sur le Lait Aigri*. Paris: Rémy, 1905.

Metchnikoff, Elie. *The New Hygiene: Three Lectures on the Prevention of Infectious Diseases.* London: William Heinemann, 1906.

Metchnikoff, Elie. *The Prolongation of Life: Optimistic Studies.* Translated from the French by P. Chalmers Mitchell. London: William Heinemann, 1907.

Metchnikoff, Elie. "Moe prebyvanie v Messine." *Russkie Vedomosti,* no. 302, December 31, 1908, 2. Reprinted in *SV,* 70–76.

Metchnikoff, Elie. "Rasskaz o tom, kak i pochemu ia poselilsia za granitsei." *Russkie Vedomosti,* no. 239, October 18, 1909, 3. Reprinted in *SV,* 77–86.

Metchnikoff, Elie. "Sur l'état actuel de la question de l'immunité dans les maladies infectieuses," 1–24. In *Les Prix Nobel en 1908.* Stockholm: Imprimerie royale, 1909.

Metchnikoff, Elie. "The Utility of Lactic Microbes." *Century Magazine* (illustrated monthly), vol. 79, November 1909, 53–58.

Metchnikoff, Elie. "Les microbes lactiques et leur utilité pour la santé." *La Revue* 2, January 15, 1911, 145–60.

Metchnikoff, Elie. "Den' u Tolstogo v Yasnoi Polyane." *Russkoe Slovo,* no. 225, September 30, 1912. Reprinted in *SV,* 128–41.

Metchnikoff, Elie. "Die Lehre von den Phagocyten und deren experimentelle Grundlagen." In *Handbuch der pathogenen Mikroorganismen,* edited by W. Kolle and A. Wassermann, vol. 2, part 1, 655–731. Jena, 1913.

Metchnikoff, Elie. "Dnevnik s zapisiami samonabliudenii." October 19, 1913–June 18, 1916. ARAN 584.2.6, 17 p.

Metchnikoff, Elie. *Sorok let iskaniia ratsional'nogo mirovozzreniia.* Moscow: Nauchnoe Slovo, 1913. Reprinted in *ASS,* vol. 13, 9–226.

Metchnikoff, Elie. "Geschichte der Phagozytenlehre." In *Handbuch der Immunitaetsforschung und experimentellen Therapie,* edited by Rudolf Kraus and Constantin Levaditi, 124–30. Jena, 1914.

Metchnikoff, Elie. *Osnovateli sovremennoi meditsiny: Paster, Lister, Kokh.* Moscow: Nauchnoe Slovo, 1915. Reprinted in *ASS,* vol. 14, 155–245.

Metchnikoff, Elie. *Etiudy o prirode cheloveka.* 5th ed. Moscow: Nauchnoe Slovo, 1917.

Metchnikoff, Elie. *Etiudy optimizma.* 4th ed. Moscow: Nauchnoe Slovo, 1917.

Metchnikoff, Elie. "Introduction." *Sur la fonction sexuelle* (unfinished book). ARAN 584.1.251, 2–10. Reprinted in *Mercure de France,* 1917; *Russkie Vedomosti,* no. 1, January 1, 1917, 4; and *ASS,* vol. 16, 283–88.

Metchnikoff, Elie. *Stranitsy vospominanii.* Moscow: USSR Academy of Sciences, 1946.

Metchnikoff, Elie. *Akademicheskoe sobranie sochinenii.* Moscow: Medgiz, 1950–1964, 16 volumes.

Metchnikoff, Elie. *Pis'ma (1863–1916).* Edited by A. E. Gaisinovich and B. V. Levshin. Moscow: Nauka, 1974.

Metchnikoff, Elie. *Pis'ma k Ol'ge Mechnikovoi, 1876–1899.* Edited by A. E. Gaisinovich and B. V. Levshin. Moscow: Nauka, 1978.

Metchnikoff, Elie. *Pis'ma k Ol'ge Mechnikovoi, 1900–1914.* Edited by A. E. Gaisinovich and B. V. Levshin. Moscow: Nauka, 1980.

Milenushkin, Iu. I. "I. I. Mechnikov i ego mesto v istorii mikrobiologii." *Mikrobiologia* 39, no. 8 (1970): 533–36.

Miterev, G. A. "Il'ia Il'ich Mechnikov." *Pravda,* May 14, 1945, 3.

Mnukhin, L. A., ed. *Russkoe zarubezh'e: Khronika nauchnoi, kul'turnoi i obshchestvennoi zhizni, Frantsiia, 1920–1940,* vol. 2. Moscow: EKSMO, 1995.

Nathan, Carl. "Metchnikoff's Legacy in 2008." *Nature Immunology* 2, no. 7 (2008): 695–98.

Needham, Joseph, with Lu Gwei-Djen. *Science and Civilisation in China.* Cambridge: Cambridge University Press, 2000.

New York Times. "Prof. Metchnikoff, Scientist, Is Dead." July 16, 1916, 17.

New York Times. "Metchnikoff," an obituary. July 17, 1916, 10.

Nikolaev, P. A., ed. *Russkie pisateli 1800–1917.* Moscow: Bol'shaia Rossiiskaia Entsiklopediia, 1999, s.v. "Nevakhovich L. N."

Oeppen, Jim, and James W. Vaupel. "Broken Limits to Life Expectancy." *Science* 296 (2002): 1029–31.

Olshansky, S. Jay, and Bruce A. Carnes. *A Measured Breath of Life.* Ebook, 2013. www.sjayolshansky.com.

Omelianskii, V. L. "Razvitie estestvoznaniia v Rossii." In *Istoriia Rossii v XIX veke,* vol. 11, chapter 5, 116–44. St. Petersburg, 1911.

Paget, Stephen. *Pasteur and After Pasteur.* London: Adam and Charles Black, 1914.

Perrot, Annick, and Maxime Schwartz. *Pasteur et ses lieutenants.* Paris: Odile Jacob, 2013.

Petrie, G. F. "The Scientific Work of Elie Metchnikoff." *Nature* 149, no. 3785 (1942): 547–48.

Petrov, R. V., and T. I. Ulyankina. "Luchshe umeret' ot nostal'gii, chem pokinut' nauku." *Vestnik Rossiiskoi Akademii Nauk* 65, no. 5 (1995): 430–42.

Pfeiffer, Richard. "Weitere Untersuchungen über das Wesen der Choleraimmunität." *Zeitschrift für Hygiene und Infektionskrankheiten* 18 (1894): 1–16.

Pipes, Richard. *Russia Under the Old Regime.* London: Penguin Books, 1995.

Podolsky, Scott. "Cultural Divergence: Elie Metchnikoff's *Bacillus bulgaricus* Therapy and His Underlying Concept of Health." *Bulletin of the History of Medicine* 72, no. 1 (1998): 1–27.

Podolsky, Scott. "The Art of Medicine: Metchnikoff and the Microbiome." *Lancet* 380, no. 9856 (2012): 1810–11.

Poidevin, Raymond, and Jacques Bariety. *Les relations franco-allemandes: 1815–1975.* Paris: Armand Colin, 1977.

Popovskii, Mark. *Vladimir Khavkin.* Jerusalem: Jews in World Culture, 1990.

Quinn, Susan. *Marie Curie: A Life.* Boston: Da Capo Press, 1995.

Rasprostranenie sifilisa v Rossii. St. Petersburg, 1895, 11 p.

Rettger, Leo F., and Harry A. Cheplin. *A Treatise on the Transformation of the Intestinal Flora with Special Reference to the Implantation of Bacillus Acidophilus.* New Haven: Yale University Press, 1921.

Reznik, Semyon. *Mechnikov.* Moscow: Molodaia Gvardia, 1973.

Riley, James C. *Rising Life Expectancy: A Global History.* Cambridge: Cambridge University Press, 2001.

Robinson, Andrew. *Sudden Genius?* Oxford: Oxford University Press, 2010.

Rousseau, Jean-Jacques. *Emile, or On Education.* Translated by Allan Bloom. New York: Basic Books, 1979.

Roux, Émile. *Pis'ma k I. I. Mechnikovu i O. N. Mechnikovoi, 1888–1914.* Moscow: Nauka, 1986.

Saada-Rémy, Elise. "Souvenirs de Lili sur son parrain." January 1970. AIP, DEL.D2, 33 p.

Schmalstieg, Frank C. Jr., and Armond S. Goldman. "Jules Bordet (1870–1961): A bridge between early and modern immunology." *Journal of Medical Biography* 17 (2009): 217–24.

Scientific American. "The Dean of Bacteriologists: A Brief Survey of the Contributions of Elie Metchnikoff to Human Welfare." Vol. 115, no. 5 (July 29, 1916): 98.

Severiukhin, D. Ia., and O. L. Leikind. *Khudozhniki russkoi emigratsii, 1917–1941.* A biographical dictionary. St. Peterburg: Izdatel'stvo Chernysheva, 1994, s.v. "Mechnikova."

Shapin, Steven, and Christopher Martyn. "How to Live Forever: Lessons of History." *British Medical Journal* 321, no. 7276 (2000): 1580–82.

Shaw, George Bernard. *The Doctor's Dilemma,* 1906. Accessed August 26, 2014. www.gutenberg.org/files/5070/5070-h/5070-h.htm.

Shevelev, A. S., and R. F. Nikolaeva. *Poslednii podvig Lui Pastera.* Moscow: Meditsina, 1988.

Shrimpton, Nicholas. "'Lane, You're a Perfect Pessimist': Pessimism and the English *Fin de siècle.*" *Yearbook of English Studies* 37, no. 1 (2007): 41–57.

Shtreikh, S. "Evreiskie otnosheniia I. I. Mechnikova." *Novyi Put',* no. 24, July 3, 1916, 7–9.

Shtreikh, S. "Iz professorskoi deiatel'nosti I. I. Mechnikova." *Russkaia Shkola,* July 5–6, 1916, 74–80.

Silverstein, Arthur M. "Introduction to the Dover Edition." In Elie Metchnikoff, *Lectures on the Comparative Pathology of Inflammation,* translated by F. A. Starling and E. H. Starling, v–ix.. Reprint, New York: Dover, 1968.

Silverstein, Arthur M. *Paul Ehrlich's Receptor Immunology: The Magnificent Obsession.* San Diego: Academic Press, 2002.

Silverstein, Arthur M. Review of *The Plato of Praed Street,* by Michael Dunnill. *Bulletin of the History of Medicine* 76 (2002): 158–60.

Silverstein, Arthur M. *A History of Immunology.* 2nd ed. Amsterdam: Academic Press/Elsevier, 2009.

Silverstein, Arthur M., and Alexander A. Bialasiewicz. "A History of Theories of Acquired Immunity." *Cellular Immunology* 51(1980): 151–67.

Silvertown, Jonathan. *The Long and the Short of It.* Chicago: University of Chicago Press, 2013.

Simmons, John Galbraith. *Doctors and Discoveries: Lives That Created Today's Medicine.* Boston: Houghton Mufflin Company, 2002.

Slosson, Edwin. *Major Prophets of Today.* Boston: Little, Brown, and Company, 1914.

Sowerine, Charles. *France Since 1870.* New York: Palgrave Macmillan, 2009.

Spafarii, N. Milesku. *Sibir' i Kitai.* Kishinev: Kartia Moldoveniaske, 1960.

Stambler, Ilia. "The Unexpected Outcomes of Anti-Aging, Rejuvenation, and Life Extension Studies: An Origin of Modern Therapies." *Rejuvenation Research* 17, no. 3 (February 2013). Accessed August 30, 2015. doi: 10.1089/rej.2013.1527.

Stambler, Ilia. *A History of Life-Extensionism in the Twentieth Century.* CreateSpace Independent Publishing, 2014.

Stefater James A., III, Shuyu Ren, Richard A. Lang, and Jeremy S. Duffield. "Metchnikoff's Policemen: Macrophages in Development, Homeostasis and Regeneration." *Trends in Molecular Medicine* 17, no. 12 (December 1, 2011): 743–52.

Strand Magazine. "Who Are the Ten Greatest Men Now Alive? A Symposium of Representative Opinions," vol. 42, no. 252, December 1911, 710–17.

Suzuki, Sarah. *The Paris of Toulouse-Lautrec.* New York: Museum of Modern Art, 2014.

Tauber, Alfred I. "The Birth of Immunology, III. Fate of the Phagocytosis Theory." *Cellular Immunology* 139 (1992): 505–30.

Tauber, Alfred I. "Ilya Metchnikoff: From Evolutionist to Immunologist and Back Again." In *Outsider Scientists: Routes to Innovation in Biology*, edited by Oren Harman and Michael R. Dietrich, 259–74. Chicago: University of Chicago Press, 2013.

Tauber, Alfred I., and Leon Chernyak. "The Birth of Immunology, II. Metchnikoff and His Critics." *Cellular Immunoloty* 121 (1989): 447–73.

Tauber, Alfred I., and Leon Chernyak. *Metchnikoff and the Origins of Immunology: From Metaphor to Theory*. New York: Oxford University Press, 1991.

Thomson, Belinda. *The Post-Impressionists*. Oxford: Phaidon Press Limited, 1983.

Times Literary Supplement. "A Naturalist on Human Nature." From a correspondent in Paris. May 22, 1903, 159–60.

Tissier, Henry, and Pascal Gasching. "Recherches sur la fermentation du lait." *Annales de l'Institut Pasteur* 17, no. 8 (1903): 540–63.

Tobin, D. J., and R. Paus. "Graying: Gerontobiology of the Hair Follicle Pigmentary Unit." *Experimental Gerontology* 36 (2001): 29–54.

Transactions of the Seventh International Congress of Hygiene and Demography. London: Eyre and Spottiswoode, 1892.

Tuchman, Barbara W. *The Guns of August*. New York: Macmillan Company, 1962.

Ulyankina, T. I. *Zarozhdenie immunologii*. Moscow: Nauka, 1994.

Vallin, Jacques. "Mortality in Europe from 1720 to 1914: Long-Term Trends and Changes in Patterns by Age and Sex." In *The Decline of Mortality in Europe*, edited by R. Schofield, D. Reher, and A. Bideau, 38–67. Oxford: Clarendon Press, 1991.

Varol, Chen, Alexander Mildner, and Steffen Jung. "Macrophages: Development and Tissue Specialization." *Annual Review of Immunology* 33 (2015): 643–75.

Vaupel, James W. "Biodemography of Human Ageing." *Nature* 464 (2010): 536–42.

Veniukov, M. I. *Iz vospominanii M. I. Veniukova*. Amsterdam, 1896.

Voltaire. "On Inoculation." October 3, 1753. Accessed May 4, 2015. www.bartleby.com/34/2/11.html.

Wawro, Geoffrey. *The Franco-Prussian War*. Cambridge: Cambridge University Press, 2003.

Weber, Eugen. *The Nationalist Revival in France, 1905–1914*. University of California Press, 1968.

Weber, Eugen. *France, Fin de Siècle*. Cambridge, MA: Belknap Press, 1986.

Wiley, Harvey W. "Sour Milk and Longevity." In *Foods and Their Adulteration*, 3rd ed., 554–57. Philadelphia: P. Blakiston's Son & Co., 1907.

Witte, Sergei. *Memoirs*. Tallin: Skif Aleks, 1994.

Wynn, Thomas A., Ajay Chawla, and Jeffrey W. Pollard. "Macrophage Biology in Development, Homeostasis and Disease." *Nature* 496 (2013): 445–55.

Zegebart, Sofia. "Vospominaniia ob I. I. Mechnikove." ARAN 584.2.18, 2 p.

Zeiss, Heinz, and Richard Bieling. *Behring: Gestalt und Werk*. Berlin-Grunewald: Bruno Schulz Verlag, 1940.

Ziegler, Ernst. *Lehrbuch der allgemeinen pathologischen Anatomie und Pathogenese*. 5th ed., vol. 1. Jena: Verlag von Gustav Fischer, 1887.

A list of Metchnikoff's publications, with links to some of the works, is provided on the website of Pasteur Institute's Médiathèque, www.pasteur.fr/infosci/biblio/ressources/histoire/metchnikoff.php.

INDEX